AIR CRASH INVESTIGATIONS

THE CRASH OF HELIOS AIRWAYS FLIGHT 522

AIR CRASH INVESTIGATIONS

Over the last decades flying has become an every day event, there is nothing special about it anymore. Safety has increased tremendously, but unfortunately accidents still happen. Every accident is a source for improvement. It is therefore essential that the precise cause or probable cause of accidents is as widely known as possible. It can not only take away fear for flying but it can also make passengers aware of unusual things during a flight and so play a role in preventing accidents.

Air Crash Investigation Reports are published by official government entities and can in principle usually be down loaded from the websites of these entities. It is however not always easy, certainly not by foreign countries, to locate the report someone is looking for. Often the reports are accompanied by numerous extensive and very technical specifications and appendices and therefore not easy readable. In this series we have streamlined the reports of a number of important accidents in aviation without compromising in any way the content of the reports in order to make the issue at stake more easily accessible for a wider public.

Hans Griffioen, editor.

AIR CRASH INVESTIGATIONS

THE CRASH OF HELIOS AIRWAYS FLIGHT 522

How crew and passengers of a Boeing 737 lost consciousness and crashed, after a long ghost flight, in the mountains of Greece, killing all 121 people aboard

Hans Griffioen, editor

MABUHAY PUBLISHING

The Crash of Helios Airways Flight 522

The details of the final Aviation Investigation Report of the Air Accident Investigation Board (AAIASB) of the Republic of Cyprus concerning the crash of Helios Airways Flight 522.

A Lulu.com imprint

ISBN: 978-1-4092-8545-8

5

Contents

PREFACE

On 14 August 2005, a Boeing 737-300 aircraft, registration number 5B-DBY, operated by Helios Airways, departed Larnaca, Cyprus at 06:07 h for Prague, Czech Republic, via Athens, Greece. As the aircraft climbed through 16.000 ft, the Captain contacted the company Operations Centre and reported a Take-off Configuration *Warning and an Equipment Cooling system problem.* Several communications between the Captain and the Operations Centre took place in the next eight minutes and ended as the aircraft climbed through 28.900 ft. Thereafter, there was no response to radio calls to the aircraft. The aircraft leveled off at FL340.

At 07:2 1 h, the aircraft flew over the Athens International Airport, and was intercepted by two F-16 aircraft of the Hellenic Air Force. One of the F- 16 pilots observed the aircraft at close range and reported at 08:32 h that the Captain's seat was vacant, the First Officer's seat was occupied by someone who was slumped over the controls, the passenger oxygen masks were seen dangling and three motionless passengers were seen seated wearing oxygen masks in the cabin. No external damage or fire was noted and the aircraft was not responding to radio calls. At 08:49 h, he reported a person entering the cockpit and occupying the Captain's seat. At 08:50 h, the left engine flamed out. and the aircraft started descending. At 08:54 h, two MAYDAY messages were recorded on the CVR.

At 09:00 h, the right engine also flamed out. The aircraft continued descending and impacted hilly terrain at 09:03 h approximately 33 km northwest of the Athens International Airport. The 115 passengers and 6 crew members on board were fatally injured. The aircraft was destroyed.

Hans Griffioen
Manila, Philippines, summer 2009

CHAPTER 1

THE BASIC FACTS

Helios Airways

Helios Airways was a low-cost airline operating scheduled and charter flights between Cyprus and many European destinations. Its main base was Larnaca International Airport, Cyprus. Flights ceased on 6 November 2006 because the Company's aircraft were detained and its bank accounts frozen by the Government of Cyprus.

The airline was established as "Helios Airways" on 23 September 1998 and was the first independent privately owned airline in Cyprus. On 15 May 2000, it operated its first charter flight to London Gatwick. It was formed by the owners of TEA (Cyprus), a Cypriot offshore air operator specialising in Boeing 737 wet leases worldwide. Originally, it offered charter services and added scheduled services on 5 April 2001. Helios was acquired in 2004 by Libra Holidays Group of Limassol, Cyprus

Helios Airways Flight 522 (HCY 522 or ZU522) was a Boeing 737-31S The aircraft involved in this incident was first flown on 29 December 1997 and had been operated by **DBA** (DBA Luftfahrtgesellschaft mbH) a low cost airline based in Munich, Germany. until it was leased by Helios Airways on 16 April 2004 and nicknamed *Olympia*, with registration 5B-DBY. At the time of the accident the aircraft was owned by Deutsche Structured Finance & Leasing GMBH & CO.

Helios Airways aircraft 5B-DBY at Luton Airport in 2004

Aside from the downed aircraft, the Helios fleet consisted of two leased Boeing 737-800s and an Airbus A319-111. delivered on 14 May 2005.

History of Flight

On 13 August 2005, on the flight prior to the accident, the Helios Airways (the Operator) Boeing 737-300 aircraft, Cyprus registration 5B-DBY, departed London-Heathrow, United Kingdom for Larnaca, Cyprus at 21:00 h. The aircraft landed at Larnaca at 01:25 h on 14 August 2005.

During the flight, the cabin crew noted a problem with the right aft service door. The cabin crew made an entry in the Aircraft Cabin Defect Log that "*Aft service door (starboard) seal around door freezes & hard bangs are heard during flt* [flight]". The write-up by the cabin crew was transferred to the Aircraft Technical Log by the flight crew as "*Aft service door requires full inspection.*"

Immediately after the arrival of the aircraft in Larnaca, the authorized company Ground Engineer (number one) went to the Boeing 737 aircraft for the required inspections. He performed a visual inspection of the aft service door and he carried out a cabin pressurization leak check. In response to the write-up in the Aircraft Technical Log, the Ground Engineer documented his actions as follows:

"*Door and local area inspected. NIL defects. Pressure run carried out to max diff. Safety valve operates at 8.25 Δpsi. No leaks or abnormal noises (IAW MM 21 – 32 – 21 – 725 – 001)*".

The aircraft was released for the next flight at 03:15 h on 14 August 2005. The aircraft was scheduled for flight HCY522, departing Larnaca at 06:00 h on 14 August 2005, via Athens, Greece to Prague, Czech Republic. The crew arrived at the company Operations Centre in Larnaca before 05:00 h and held a briefing for the flight.

The aircraft took off from Larnaca airport at 06:07:13 h. At 06:11:21 h, the flight crew contacted Nicosia Area Control Centre (ACC) at reporting point LOSOS, climbing through flight level (FL) 100 for FL200. At 06:11:35 h, Nicosia ACC identified the flight and asked for the requested final cruising level. The Captain requested cruising level 340.

At 06:11:45 h, flight HCY522 was cleared to climb to FL340 and to proceed direct to the RDS (Rodos) VOR. The Captain acknowledged the clearance. This was the last recorded communication between the flight crew and Nicosia ACC.

At 06:12 h, Nicosia ACC called the Planner Controller in Athinai ACC with data on flight HCY522: FL340 and 06:37 h estimated time at reporting point EVENO between the Athinai and Nicosia FIRs. The Planner Controller verified the data, which had been received earlier, as an "estimated message" via Aeronautical Fixed Telecommunication Network (AFTN).

According to the Flight Data Recorder (FDR), at 06:12:38 h and at an aircraft altitude of 12 040 ft and climbing, the cabin altitude warning horn sounded.

At 06:14:11 h, at an aircraft altitude of 15.966 ft, the Captain contacted the company Operations Centre on the company radio frequency, 131.2 MHz. According to the Operator's Dispatcher, the Captain reported:

"Take-off configuration warning on" and *"Cooling equipment normal and alternate off line."*

The Dispatcher requested an on-duty company Ground Engineer to communicate with the Captain.

According to a written statement by the Ground Engineer (number one), written immediately after the accident at the Technical Manager's instruction, the Captain reported that *"the ventilation cooling fan lights were off."* Due to the lack of clarity in the message, the Ground Engineer asked him to repeat. Then, the Captain replied *"where are the cooling fan circuit breakers?"* The Ground Engineer replied *"behind the Captain's seat."*

According to another statement given by the Ground Engineer to the Cyprus Police on 19 August 2005, the Captain reported *"both my equipment cooling lights are off."* The Ground Engineer replied *"this is normal"* and asked the Captain to confirm the problem *"because it did not make sense, as the lights are normally off when the system is serviceable"* [i.e. operating properly]. The Captain replied *"they are not switched off."* Since the message from the Captain did not make any sense to the Ground Engineer and *"given the close proximity of the pressure control panel and the fact that he* [the Ground Engineer] *had used the pressure panel prior to the flight and the pressure panel has four lights"*, the Ground Engineer asked the Captain to *"confirm that the pressurization panel was selected to AUTO."* The Captain replied *"where are my equipment cooling circuit breakers?"* The Ground Engineer replied *"behind the Captain's seat".*

During the communication between the flight crew and the company Operations Centre, the passenger oxygen masks deployed in the cabin as they were designed to do when the cabin altitude exceeded 14 000 ft. It was determined that the passenger oxygen masks deployed at 06:14 h at an aircraft altitude of approximately 18 000 ft (extrapolation of the data from the NVM in the cabin pressure controller).

According to the FDR, the microphone keying (communication between the Captain and the Ground Engineer) ended at 06:20:21 h as flight HCY522 was passing through 28 900 ft. Shortly afterwards, the Operator's Dispatcher called the flight crew again but there was no response.

At 06:23:32 h, the aircraft leveled off at FL340.

At 06:29 h, the Operator's Dispatcher called Nicosia ACC and asked the Air Traffic Controller to contact flight HCY522.

From 06:30:40 to 06:34:44 h, Nicosia ACC called flight HCY522 without receiving any response. At 06:35:10 h, Nicosia ACC tried to make contact with the flight via another aircraft without any response.

At 06:35:49 h, Nicosia ACC called the flight requesting it to "Squawk STAND-BY."

At 06:36:00 h, one minute before flight HCY522 entered the Athinai FIR, the color of its radar track and the corresponding label on the radar display in the Athinai ACC, changed automatically from green to salmon color. The color of the target and the label then changed to blue when the Athinai Controller "clicked" on the radar target, which meant that he was aware of the incoming flight and that automatic coupling of track and Flight Plan had occurred.

At 06:36:12 h, Nicosia ACC contacted the Planner Controller at Athinai ACC with the information that flight HCY522 was *"over point EVENO* [entry point for the Athinai FIR] *and does not answer, if he calls you, let us know."*

At 06:37:27 h, the flight entered the Athinai FIR, about 10 NM south of point EVENO without calling Athinai ACC. The flight continued at FL340 towards Athens according to its Flight Plan route direct to RDS VOR, then via UL995 – RIPLI – VARIX – KEA VOR.

A further radio call was made by Nicosia ACC at 06:39:30 h on the emergency frequency (121.5 MHz) but there was no response from flight HCY522. At 06:40:15 h, Nicosia ACC called the Planner Controller in Athinai ACC asking *"Did Helios call you?"* The latter answered *"not yet."*

At 07:12:05 h, the Athinai ACC Radar Controller called flight HCY522 in order to issue a descent clearance. There was no response from HCY522. Further attempts to call the flight were made on the emergency frequency and by other aircraft. At 07:12:32 h, the Planner Controller in Athinai ACC called Athinai Approach Control (APP) and informed them that he had no radio contact with flight HCY522.

Between 07:12:38 h and 07:12:50 h, the Athinai ACC Radar Controller called flight HCY522 three times on the frequency in use (124.475 MHz).

At 07:12:48 h, flight HCY522 passed reporting point RIPLI towards point VARIX.

Between 07:13:04 h and 07:14:36 h, the Athinai ACC Radar Controller called flight HCY522 eleven times on the frequency in use. Between 07:12:52 h and 07:49:18 h, he called on the emergency frequency (121.5 MHz) five times for radio check, three times to *"Squawk IDENT"* and one time asking the flight to call the Athinai APP frequency. Another aircraft also called the flight on the frequency in use and on the emergency frequency.

At 07:15:19 h, the flight passed reporting point VARIX and then followed the VARIX 2C STAR to KEA VOR.

At 07:16 h, the Athinai ACC Radar Controller informed the Athinai ACC Supervisor about the radio communication failure (RCF) of flight HCY522. The Supervisor notified Athinai APP, Athens Tower and the Hellenic Air Force.

At 07:20:59 h, the flight passed the KEA VOR, and began what appeared to be a standard instrument approach procedure for landing at Athens International Airport, runway 03L, but remained at FL340. At 07:29 h, flight HCY522 flew over the Athens International Airport still at FL340 and following the missed approach procedure for runway 03L turned right towards the KEA VOR. At 07:37:39 h, flight HCY522 reached the KEA VOR and entered the published holding pattern.

At 07:53:50 h, Athinai ACC declared an ALERT phase to the Joint Rescue Coordination Center (JRCC).

At 08:23:5 1 h, during the sixth holding pattern, flight HCY522 was intercepted by two F-16 fighter aircraft of the Hellenic Air Force. The F-

16s made close visual contact with the flight in the holding pattern, at FL340. During the interception, the F16s communicated on the military radar frequency and with Athinai ACC. One of the F-16 pilots attempted to attract the attention of the flight crew using prescribed interception signals and radio calls on the emergency and Athinai ACC frequencies, without success. He maneuvered around the aircraft to acquire various views from the right and left sides of the cockpit and the fuselage in an effort to identify the reasons for the lack of radio communication. No external structural damage or fire/smoke was observed.

At 08:32 h, the F- 16 pilot reported by radio that the Captain's seat was vacant. The First Officer's seat was occupied by someone who was slumped over the controls. Two passengers on the left side of the aircraft, one wearing white clothing and the other red clothing, sat motionless in their seats and were wearing oxygen masks on their faces. Additional oxygen masks could be seen dangling from their overhead units. The passenger cabin was dark, but the shadow of the oxygen hoses and masks could be seen against the daylight shining through the windows on the other side of the passenger cabin. Another passenger was seen from the right side of the aircraft wearing white clothing, sat motionless and wore an oxygen mask.

At 08:34:00 h, Athinai ACC declared a DISTRESS phase to the JRCC.

At 08:48:31 h, two chimes were heard on the CVR and, at 08:48:51 h, another two chimes were heard followed after 20 seconds by a continuous chime which lasted 20 seconds. Some seconds later, a click sound similar to the cockpit door opening was recorded. Also, sounds similar to movement in the cockpit, seat adjustment, and oxygen mask removal from its stowage box and oxygen flow during donning of the mask were recorded.

Approximately 08:49 h, during the tenth holding pattern, the F-16 pilot observed a person wearing a light blue shirt and dark vest, but not wearing an oxygen mask, enter the cockpit and sit down in the Captain's seat. He put on a set of headphones and appeared to place his hands on the panel directly in front of him.

According to the FDR, at 08:49:50 h, the left engine flamed out. At this time, the F-16 pilot observed what he assumed was fuel coming out of the left engine. The aircraft turned steeply to the left and headed in a northerly direction. The person in the Captain's seat did not respond to any of the attempts of the F-16 pilot to attract his attention. He appeared to be bending forward every now and then. Flight HCY522 began a descent on a northwesterly heading. The two F-16s followed at a distance due to the maneuvering by the Boeing 737.

When the F-16 pilot next came close to the Boeing 737, he saw the upper body of the person in the First Officer's seat lean backwards as if he was sitting up. It became evident that this person was not wearing an oxygen mask and remained motionless.

At 08:54:18 h, the following distress was recorded by the CVR *"MAYDAY, MAYDAY, MAYDAY, Helios Airways Flight 522 Athens ... (unintelligible word)"*. A few seconds later, another *"MAYDAY, MAYDAY"* with a very weak voice was recorded.

When the Boeing 737 was at about 7.000 ft, the person in the Captain's seat for the first time appeared to acknowledge the presence of the F-16s and he made a hand motion. The F-16 pilot responded with a hand signal for the person to follow him on down towards the airport. The person in the Captain's seat only pointed downwards but did not follow the F-16.

At 08:59:20 h, the heading of the Boeing 737 changed to a southwesterly direction. The aircraft continued to descend. At 08:59:47 h, according to the FDR, the right engine flamed out at an altitude of 7.084 ft.

The aircraft continued to descend rapidly and collided with rolling hilly terrain in the vicinity of Grammatiko village, approximately 33 km northwest of the Athens International Airport at 09:03:32 h.

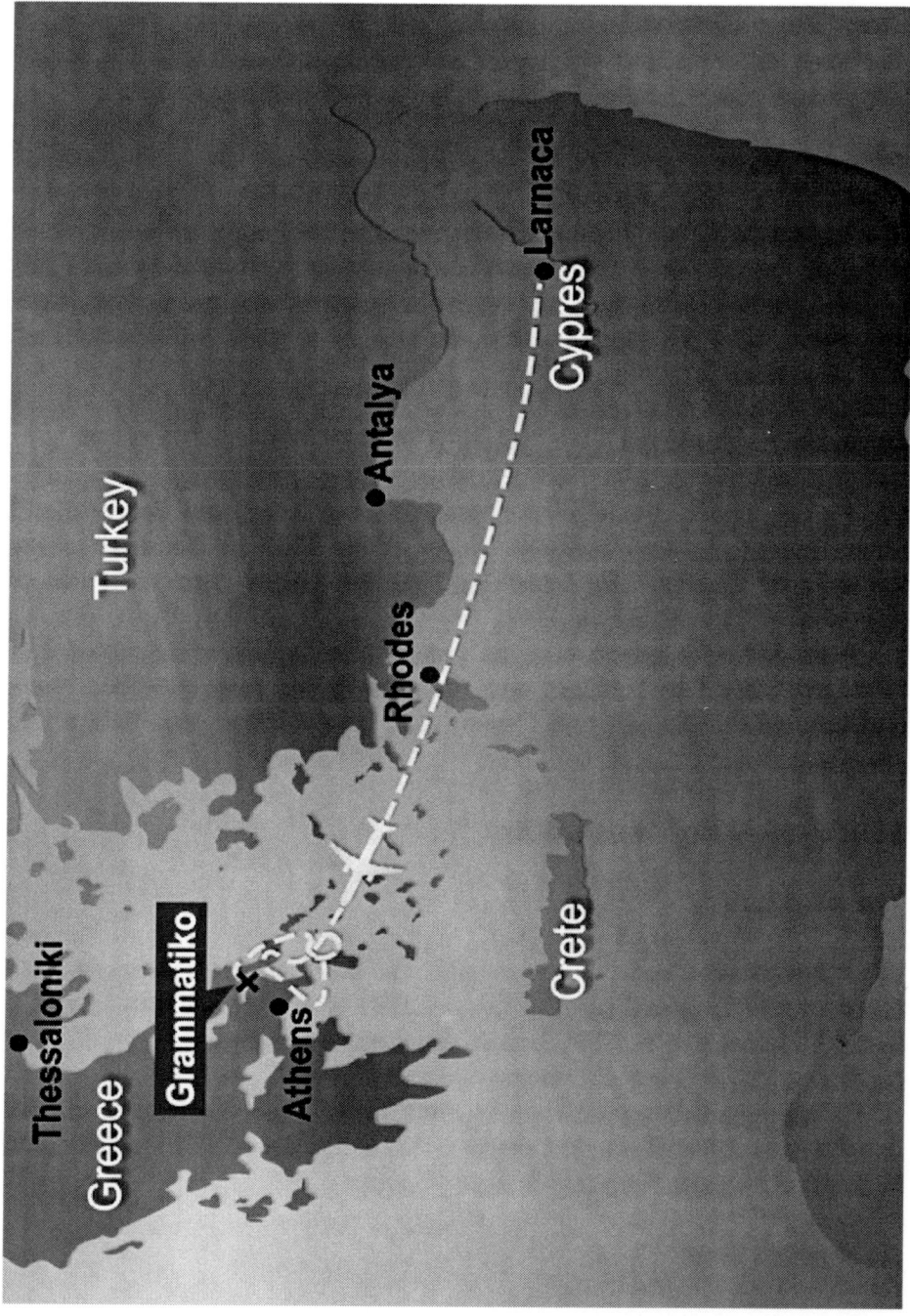

Damage to Aircraft

The aircraft was destroyed by impact forces and post impact fire.

Other Damage

The impact and post impact fire destroyed several acres of scrub brush and open pasture land. In a scientific soil analysis performed by ENVAR, localized contamination by total petroleum hydrocarbons and hydrocarbons was found. ENVAR suggested that the area be aerated and re-examined after four months.

Injuries to Persons

Of the one hundred and fifteen passengers on board, one hundred and eleven passengers were nationals either of the Hellenic Republic or the Republic of Cyprus. The remaining four passengers were nationals of Australia.

There was a six person crew on board. The Captain was a national of Germany. The First Officer and the four cabin crew members were nationals of the Republic of Cyprus. The First Officer also held a UK passport.

Personnel Information

The Captain

The Captain was male, 59 years old. He held an Air Transport Pilot License (ATPL) issued on 27 February 1991 in accordance with JAR-FCL by LBA Germany. His ATPL license, his instrument rating category III, and his Boeing 737-300 and -800 ratings were valid until 4 June 2006.

He had attended engineering college in Dresden, East Germany from 1966 to 1970 and graduated as a pilot-engineer. He had held an ATPL issued in 1970 by the Civil Aviation Authority in East Germany.

Employment history:

1970 – 1978	Interflug, First Officer
1978 – 1991	Interflug, Pilot-in-Command
1992 – 1994	Saarland Airlines, Pilot-in-Command (trained at Maersk Air in Denmark)
1996 - 1997	Air Berlin, Pilot-in-Command
1997 – 1998	Virgin Express, Belgium, Pilot-in-Command
1998 – 1999	Falcon Air, Sweden, Pilot-in-Command
1999 – 2000	Sabena Airline, Belgium, Pilot-in-Command
2000 – 2001	EasyJet, United Kingdom, Pilot-in-Command
Mar 2002 – Mar 2003	Bluebird Cargo, Iceland, Pilot-in-Command on B737-300
May 2004 – Oct 2004	Helios Airways, Cyprus, Pilot-in-Command on B737-300 and -800
Dec 2004 – Jan 2005	Travel Service, Pilot-in-Command on B737-400, -500 and -800
Mar 2005 – May 2005	Channel Express, United Kingdom, Pilot-in-Command on B737-300
May 2005 –	Helios Airways, Pilot-in-Command on B737-300 and -800

Medical Certificate: Class A, Medical Certificate issued on 21 March 2005 and valid until 9 October 2005 with the restriction to carry two pairs of corrective lenses. Last LPC/OPC4 June 2005. Recurrent Training in STD 4 June 2005. Last Line Check 12 June 2005 CRM training 2 June 2005.

Flying Experience:

- Total on all types: 16.900 h
- Total on type B737: 5.500 as Pilot in Command
- Total last 7 days duty time: 37:20 h
- Last 24 hours: Off duty
- Rest period: Off duty for 2 days
- Activities last 24 hours: Undetermined

Company limits:

Flying hours: 100 in 28 days; 900 h in 12 months.

Duty time: 55 h in a week; 95 h in 2 weeks; 190 h in 4 weeks.

The Captain had worked for the Operator for two separate time periods. According to interviews of his peers at the Operator, during the first period, he presented a typical *"command"* attitude and his orders to the First Officers were in command tone. During the second period, his attitude had improved as far as his communication skills were concerned.
According to an oral statement by the next of kin, the Captain was a quiet and professional pilot. His hobby was to construct and fly model aircraft. He used no drugs or medication, and he used alcohol occasionally and with moderation.

First Officer

The First Officer was male, 51 years old. He held an Air Transport Pilot License (ATPL) issued in accordance with JAR-FCL by United Kingdom. His ATPL license and Boeing 737-300 and -800 ratings were valid until 31 March 2006 and his instrument rating category III was valid until 31 October 2005. He had attended and graduated from Chelsea College with an Engineering Diploma. He was also trained at Oxford Air Training School to become a pilot.

Employment history:

Feb 1986 – Mar 1988	Cyprus Airways, as Ground Engineer / Line Maintenance (on B707, BAC 111, Airbus A3 10, A320)
July 1989 – Apr 1997	Hunting Cargo, as Senior First Officer (SFO)
Mar 1998 – Oct 1999	Air Scandic, as SFO
Apr 2000 – Aug 2005	Helios Airways, as SFO

Medical Certificate: Class A Medical Certificate issued on 25 April 2005 and valid until 29 October 2005 with no restrictions. Last OPC 9 March

2005. Last Line Check 3 February 2005. Recurrent training in STD
9 March 2005. CRM training 28 February 2005.

Flying Experience:

- Total on all types: 7.549 h
- Total on type B737/300-900: 3.991 h
- Total last 7 days duty time: 26:25 h
- Last 24 hours: Off duty
- Rest period: Off duty for 2 days

The First Officer spent the day preceding the accident at his summer
house with the family. He drove home in the evening, had a normal
dinner (no alcohol) and he went to bed at about 23:00 h. He woke up early
in the morning and drove to the airport in order to report for duty on
time.

According to statements by his next of kin, colleagues and friends, the
First Officer was an optimist, calm, active and a social person. He had
expressed his views several times about the Captain's attitude. He had also
complained about the organizational structure of the Operator, flight
scheduling and he was seeking another job. He used no drugs or medication
and he did not smoke or drink alcohol. In his last three OPCs, there were
the following remarks/ recommendations:

- 9 March 2005:
  *"Standards achieved, but with room for lots of improvement. Some
  difficulties met in complex tasks. Do not rush through check lists.
  Recommendation – improve your understanding on the use of AFS".*

- 3 September 2004:
  *"Overall standard is above average. Very Good LVO recognition of
  abnormalities. EMERGENCY DESCENT* [capital letters
  used by the TRE] *repeated at very good standard. Keep the good
  work".*

- 13 April 2004:

"Overall performance at standard – Good Manual control, and 1 ENG G/A. –Make positive control is advisory after engine failure on T/OFF not to loose direction. Repeat – OK".

In addition, the First Officer's training records were reviewed for the five years he worked for the Operator. The review disclosed numerous remarks and recommendations made by training and check pilots referring to checklist discipline and procedural (SOP) difficulties.

Cabin Attendants

There were four cabin crew members on board, all of which met Operator proficiency and medical requirements. All four had recently participated in the following training courses:

- Emergency procedures
- CRM
- Security
- First Aid
- Dangerous Goods Training

One cabin attendant also held a UK Commercial Pilot License (JAR CPL A/IR) with an issue date of 2 October 2003, and valid to 1 October 2008. His JAA Class 1 Medical Certificate was valid from 15 July 2005 to 17 July 2006.

Ground Engineers

Engineer number one:

Male, age 44. He held a valid JAR 66 Category B1 + B2 + C license issued on 5 July 2004 and covering all B737 aircraft types and systems. He reported that he had held a UK BCAR aircraft maintenance engineer license for B737 aircraft since 1987. He also held licenses for numerous other aircraft types. He held a valid ATC LASHAM Authorization issued on 4 May 2005 for certifying Helios/LCA B737-300 and -800 aircraft (No. ATCL 5006). He was hired by the Operator, for a second time, as a freelance engineer through a UK Agency, for the period April to September 2005 (Peak Period).

Ground Engineer number two:

Male, age 39. He held a valid UK CAA Category A Airframe and Category C Engine Aircraft Maintenance Engineer issued in October1990. He also held a valid ATC Lasham Authorization issued on 2 December 2004 for certifying B737-300 aircraft. He was employed by ATC Lasham, and was dispatched to the Operator for six weeks to support line maintenance for this period (return on 14 August 2005).

Ground Engineer number three:

Male, age 29. He held a valid UK CAA BCAR Section L Category A Airframe and Category C Engine Aircraft Maintenance Engineer license issued on 28 August 2002 and a valid ATC Lasham Authorization issued on 21 July 2005 for certifying Helios/LCA B737-800 aircraft. He was employed by the Operator in January 2001. In the summer of 2002, he joined Cyprus Airways. On 1 August 2005, he returned to work for the Operator.

Air Traffic Controllers

Nicosia ACC

The Controllers on duty at the West Sector of Nicosia ACC, two Executive Radar Controllers and two Planners, all held the required licenses and ratings, and had the required experience to execute their duties at their work positions. The two Executive Controllers were employed in 1992 and the two Planners in 1997. All had Air Traffic Controller Licenses with valid ratings in Area Control Radar and Area Procedural Control; the two Executive Radar Controllers since 1996 and the two Planners since 2000.

Athinai ACC

The Controllers on duty at the Sector of Athinai ACC involved held all the required licenses and ratings, and had the required experience to execute their duties at their work positions. The two Executive Radar Controllers held Air Traffic Controller Licenses, and valid ratings for Area Procedural Control since 1984. They held Area Radar Control ratings since 1991 and had 15 years of experience. They were instructors in Area Procedural Control at the Hellenic Civil Aviation Administration (HCAA) Training School, as well as in Area Control by Radar (classroom, simulator, and on-the-job training).

The two Planner Controllers were employed in 1999. They obtained their Air Traffic Controller Licenses and the Area Procedural Control ratings on 31 May 2001 and 18 January 2005, respectively. The second Planner also held Airport Control and Approach Control ratings.

The Supervisor held the required licenses and ratings for Procedural and Radar Controller, with a work experience of 28 years in the Athinai ACC. He was also an instructor at the HCAA. During his shift, he was responsible for monitoring and ensuring the normal and smooth operations of Athinai and Makedonia ACCs, in accordance with the national and international rules and procedures.

Aircraft Information

General

The Boeing 737-31S (B737-300 series) aircraft was built in 1998, with serial number 29099, line number 2982. The aircraft was delivered to Deutsche British Airways (DBA) on 15 January 1998, and it was registered in Germany as D-ADBQ. The Operator acquired the aircraft on 16 April 2004 and it was registered in the Republic of Cyprus as 5B-DBY. The owner of the aircraft was Deutsche Structured Finance & Leasing GMBH & Co. Cyprus DCA issued the Certificate of Airworthiness No. 244 valid from 16 April 2004 to 15 April 2006.

According to the technical records, the aircraft had accumulated 17.900:5 6 flight hours and 16.085 cycles up to 14 August 2005. The aircraft was fitted with two CFMI 56-3C1 engines; serial number 858752 in the number one position with 16.477 total hours since new; and serial number 858755 in the number two position with 16.280 total hours since new.

The aircraft was configured with 142 economy class seats. There were also four seats in the cockpit (two for the pilots and two for observers) and four cabin crew seats in the cabin (two at the front and two at the rear).

The maximum take-off mass for the aircraft was 63.276 kg and the aircraft mass at take-off was 51.572 kg. The CG range for take-off was between 5 % and 26 % Mean Aerodynamic Chord (MAC) and the center of gravity (CG) at take-off was 21.0 % MAC. Aircraft mass at the time of the accident was calculated to be approximately 45.000 kg, and the MAC was calculated to be approximately 23.5 %. The aircraft was loaded with 7.000 kg of Jet A1 type fuel. No deferred defect items were pending in the technical logbook. The aircraft's certified ceiling was 37.000 ft.

Regarding aircraft manuals, the Board asked Boeing whether the Operator had a contract for receiving issued amendments to the following manuals: FCOM, QRH, AFM, and AMM. Boeing responded as follows:

"For the 737-300, Helios Airways (HCY) was receiving the Airplane Flight Manual (AFM) and the Aircraft Maintenance Manual (AMM) from Boeing. At HCY's request, Boeing shipped three copies of the Deutsche Structured Finance GMBH (DBA) Flight Crew Operations Manual (FCOM) and Quick Reference Handbook (QRH) manuals to HCY on 17 June 2004. HCY did not contract with Boeing for its own customized versions of the FCOM or QRH manuals or revisions of those DBA manuals. DBA was receiving revisions of the FCOM and QRH from Boeing.

For the 737-800, HCY does have a contract with Boeing for the FCOM, QRH, AFM, and AMM and the issuance of amendments."

The Board established that DBA continued to receive the revisions and amendments from Boeing for the relevant manuals applicable to the B737-300 aircraft now operated by the Operator, and apparently the Operator received the relevant revisions from DBA. The B737-300 Operations Manual, Volume 1 contained the Temporary Revision No. 2 dated 25 July 2005, the B737-300 Operations Manual, Volume 2 contained the Temporary Revision No. 1 dated 24 January 2005, and the B737-300 Quick Reference Handbook (QRH) contained revisions dated 3 June 2005. However, the Board was informed that any revisions to the manuals received through DBA were evaluated by the staff of the

Operator and all revisions were not necessarily incorporated in the Operator's Manuals. In this respect, the Board noted that the older versions of the After Takeoff checklists in the FCOM and the QRH contained an item *"AIR COND & PRESS ..... SET"* and in the revised FCOM and QRH this item had been changed to two items *"Engine bleeds ..... ON"* and *"Packs ..... AUTO"*. However, this change had not been incorporated by the Operator in the flight crew manuals.

Maintenance History

The aircraft was maintained in accordance with the B 737-3 00 Maintenance Program, reference DCA MS 003/2004, based on Boeing's MPD (Maintenance Planning Document), reference D6-38278, date November 1988, revision level January 2004, approved on 19 July 2005 by the Cyprus Department of Civil Aviation in compliance with the requirements of JAR AMC - OPS 1.910(a). In accordance with these requirements, a Certificate of Release to Service (CRS) was issued after every inspection check in accordance with the Maintenance Program and signed by authorized maintenance personnel.

Scheduled Maintenance

The aircraft was delivered to the Operator on 4 April 2004 with 14.591 flight hours and 14.365 cycles (flights) since new. As of this date and up to the day of the accident, 15 checks had been performed in accordance with the approved maintenance program. The checks are based on documented information gathered from the Aircraft Technical Log and ATC Lasham maintenance records.

Unscheduled Maintenance

Decompression Occurrence on 16 December 2004. According to Technical Log entries, unscheduled maintenance tasks were performed at Larnaca on 19 December 2004 by an authorized Ground Engineer of ATC Lasham, in order to investigate the cause of a rapid decompression which had occurred on 16 December 2004. It should be noted that this event occurred just 07:31 flight hours and seven landings after a "C" check performed on 10 December 2004.

Maintenance action was taken by the authorized Ground Engineer at Larnaca airport in response to the pilot write-up in the Aircraft Technical Log (p. 650): "*Rapid decompression. Descent to 10 000. Pax O$_2$ deployed.*"

As a result of the maintenance action taken, the Ground Engineer recorded the following corrective actions: "*Aft R/H service door adjusted and rigged acc to MM adjustments and test procedure 52-41-00.*"

The same day, the maintenance management requested removal of the CVR and FDR for downloading (p. 652), and change of the cabin pressure controller No. 2 (p. 653). All requests were accomplished. Finally, according to p. 654 of the Aircraft Technical Log, a cabin pressure leak check was carried out as per AMM 05-51-91-702-001 and an outflow valve operational test as per AMM 21-31-11 -7-15-051. No abnormal findings were recorded.

On 19 December 2004, the aircraft was ferry flown to ATC Lasham with the passenger cabin oxygen system deployed and the oxygen generators depleted. During maintenance at ATC Lasham, the oxygen generators were replaced. No significant findings were noted. On 22 December 2004, the aircraft was flown to Larnaca and returned to service.

Aft Service Door Occurrence on 13 August 2005. The following information is based on three statements submitted by Ground Engineer number one. The first statement (dated 14 August 2005) was a report written on the day of the accident on a laptop in the company office (the Technical Manager told the Board that upon being notified of the accident, he had directed the Ground Engineer to write a report about the maintenance actions that he had carried out on the aircraft that morning). The second statement (dated 19 August 2005) was a statement given to the Cyprus Police. The third statement (dated 12 September 2005) was given to the Board in the course of the investigation at a meeting that took place in the United Kingdom.

In the evening of 13 August 2005, Ground Engineer number one, who was to commence his duty at 01:30 h on 14 August, had agreed to provide Ground Engineer number two transportation to the airport, as the latter was going to be catching a plane to return home (London). The two men arrived at the airport at approximately 01:15 h. Ground Engineer number two remained in the company's Engineering Office to wait for confirmation whether he could have a supernumerary crew seat on a company flight to Luton, UK that morning. Ground Engineer number one went to meet the incoming flight.

The aircraft arrived at Larnaca, Cyprus at 01:25 h in the early morning of 14 August 2005. During the flight from London to Larnaca the cabin crew wrote in the Cabin Log that *"Aft Service Door (starboard) seal around door freezes & hard bangs are heard during flt."* The statement was transferred to the Aircraft Technical Log by the flight crew who stated that the *"Aft service door requires full inspection."* Ground Engineer number one reviewed the entries in the Technical Log and determined that an aft service door visual inspection and a cabin pressurization leak check would be required. He realized that he would need some help in performing the leak check. Therefore, he asked Ground Engineer number two to assist him in a *"pressure run/test"*. Another company engineer (Ground Engineer number three, authorized only on the B737-800 aircraft, was awaiting the arrival of that aircraft) assisted as safety man positioned outside the aircraft and to listen to any sounds of air leaking from the aircraft.

According to Ground Engineer number one, before going to the aircraft to perform the maintenance actions, he had printed from the company computer a paper copy of the AMM procedure/task for the test he was going to perform. He stated that he brought no stopwatch (or other equipment) with him to the aircraft, as *"it was not required for the test."*

A visual inspection of the rear right door (R2) was carried out from the inside, as well as from the outside using a catering truck for access. The door was opened and closed several times. It operated normally and no defects were found.

For the cabin pressurization test, Ground Engineer number two stayed in the rear of the aircraft by the R2 door, while Ground Engineer number one went to the flight deck. He pressurized the aircraft using the APU and both air conditioning. He stated that he *"selected the press controller to MAN and selected full closed"*, and that *"The air conditioning discharge valve closed at about 2 psi."* The air conditioning packs were selected one at a time in AUTO and HIGH flow, and were observed to be working correctly. *"At about 5 psi,"* Ground Engineer number two experienced some ear pain, walked up to the flight deck and asked Ground Engineer number one to slow down the rate of pressurization. Ground Engineer number two then returned to the aft cabin to listen for sounds of leakage, while he remained in contact with Ground Engineer number one on the flight deck through the intercom system.

When Ground Engineer number two indicated it was ok to do so, Ground Engineer number one increased the cabin differential pressure *"to 8.25 psi"* where *"the safety valve started operating."* At that time, Ground Engineer number one left the flight deck and joined Ground Engineer number two in the aft cabin listening for sounds of leakage inside of the aircraft at the rear right door. When satisfied that everything was normal, he returned to the flight deck. This condition (8.25 Δpsi) was maintained for about 5 minutes. Ground Engineer number one then started depressurizing the aircraft. At about 5 Δpsi, both air conditioning packs were selected to OFF. At about 4.5 Δpsi, the rate of descent was 3500 ft/min and the air conditioning packs were selected ON to slow the depressurization rate (due to the earlier ear problem of Ground Engineer number two).

According to Ground Engineer number one, when the aircraft was depressurized to approximately 0.5 Δpsi, he selected the pressurization mode selector to AUTO and *"as the cabin VSI* [vertical speed indicator] *started to decrease, opened the DV* [Direct Vision - flight deck sliding window]." Ground Engineer number two then went to open the cabin doors, while Ground Engineer number three left to meet the B737-800 aircraft that had arrived. Ground Engineer number one remained on the flight deck and carried out an in situ (i.e., without removing them from their stowed position) flight crew oxygen mask function test by *"pressing the oxygen test button and audio testing."* The AMM procedure/task being followed for the Cabin Pressure Leak Test prescribed such an oxygen mask test if the absolute pressure reached during the leak test exceeded 20 psi. Ground Engineer number one stated that the pressure in the aircraft had probably reached about 23 psia.

Ground Engineer number two waited outside the aircraft in a car for Ground Engineer number one. They then drove back to the office, where, according to the statement given by Ground Engineer number one, he *"consulted the Maintenance Manual on the computer to confirm the safety valve operation limits."* He then documented the maintenance actions in the Aircraft Technical Log by making four separate entries on two separate pages of the logbook. The first, third, and fourth entry referred to routine maintenance actions (daily check, flight deck door compression latch check, and QAR memory card replacement). The second entry was in response to the write up by the flight crew in the Aircraft Technical Log regarding the aft service door and referred to the pressurization test that had just been performed. Ground Engineer number one wrote: *"Door and local area inspected nil defects. Pressure run carried out to max diff. Safety valve operates at 8.25 Δpsi. No leaks or abnormal noises IA W MM 21-32-725-001."*

According to the Technical Log entry, the daily check was accomplished and signed off at 02:30 h and the rest of the tests were accomplished and signed off at 03:15 h.

Systems Descriptions

Bleed Air System

Air for the bleed air system can be supplied by the engines, the auxiliary power unit (APU), or by an external air cart/source. The APU or external cart supplies air to the bleed air duct prior to engine start. After engine start, air for the bleed air system is normally supplied by the engines. Both bleed air switches are located on the forward overhead pressurization panel. The following systems rely on the bleed air system for operation:

- Air conditioning /pressurization

- Wing and engine thermal anti-icing

- Engine starting

- Hydraulic reservoirs pressurization

- Water tank pressurization

Air Conditioning

Conditioned air for the cabin comes from either the aircraft air conditioning system or an external ground source. The air conditioning system provides temperature-controlled air by processing bleed air from the engines, APU, or a ground air source, in the air conditioning packs. The flow of bleed air from the main bleed air duct through each air conditioning pack is controlled by the respective pack valve. Normally the left pack uses bleed air from engine No. 1 (left engine) and the right pack uses bleed air from engine No. 2 (right engine). A single pack is capable of maintaining pressurization and an acceptable temperature throughout the aircraft up to the maximum certified ceiling.

Pressurization System

The purpose of the pressurization system is to provide a safe comfortable cabin altitude at all aircraft altitudes. The air conditioning system provides a constant flow of pressurized, conditioned air to the cabin.

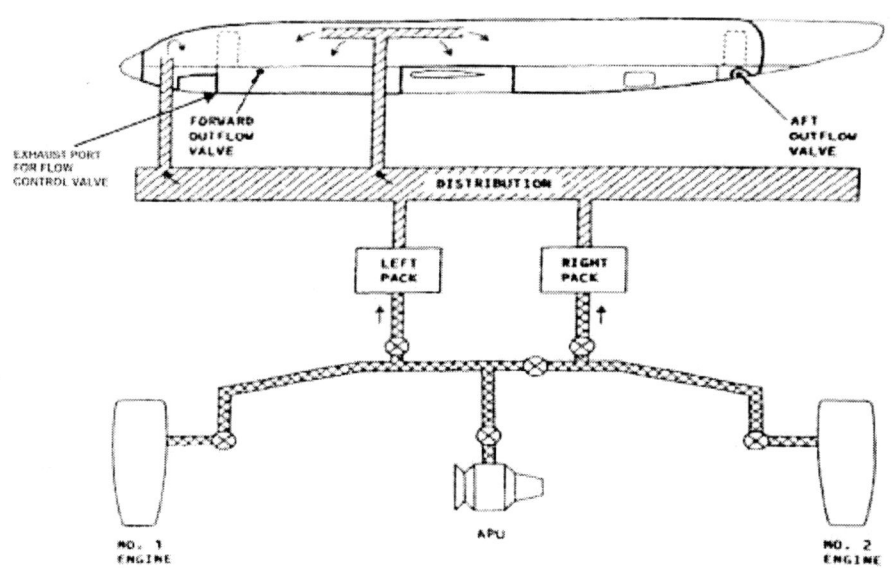

Diagram – Pressurization System

Normally, a small amount of the air leaks overboard through door seals and other openings. The pressurization system meters the amount of air exhausted from the aft outflow valve to control pressurization to a prescribed schedule depending on the aircraft altitude, cruise altitude, and the altitude of the destination.

Cabin pressurization is controlled during all phases of aircraft operation by the Digital Cabin Pressurization Control System (DCPCS). The system uses bleed air supplied by the engines and distributed to the air conditioning system. Pressurization and ventilation is controlled by modulating the outflow valve and the overboard exhaust valve. (See Diagram – Pressurization System).

Cabin altitude is normally rate-controlled by the digital cabin pressure controller up to a cabin altitude of 8 000 ft at the aeroplane maximum certified ceiling of 37 000 ft.

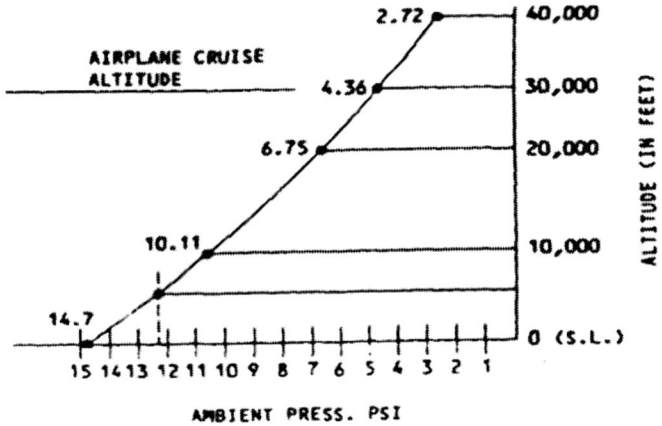

Diagram – Altitude vs Ambient Pressure

The DCPCS consists of a pressure control panel, two identical digital cabin pressure controllers, and a pressurization outflow valve. The DCPCS operates in two main modes, automatic and manual, which are selected in the flight deck using the selector switch on the pressure control panel.

There are three allowable selector switch positions: AUTO, ALTN and MAN. The AUTO and ALTN positions will provide automatic cabin pressure control, and the MAN position will allow manual control. Normally, the selector switch is positioned to AUTO.

In automatic mode, one pressure controller is operating, while the other controller is in standby mode. The DCPCS operates in the automatic mode using commands from the operational pressure controller to increase or decrease the aft outflow valve's exhaust Dareas as required during the flight to maintain the cabin pressure schedule.

- AUTO – Automatic pressurization control, normal mode of operation.

- ALTN – With the pressure control panel selector switch in the ALTN position, the DCPCS operates in the automatic mode except that the pressure controllers will be forced to swap control status, provided both are functional. The operational controller will go into standby, and the controller that was in standby will become the operational controller. The new operational pressure controller will take over the cabin pressure scheduling duties and command of the pressurization outflow valve.

- MAN – With the pressure control panel selector switch in the MAN (manual) position the DCPCS operates in manual mode. Manual mode provides direct control of the pressurization outflow valve through a separate DC manual motor on the outflow valve. The motor is controlled by the three position toggle switch on the pressure control panel. The pressurization outflow valve position relative to the full open or full closed state is displayed in the gauge directly above the toggle switch. Manual control is primarily used as a backup to automatic control.

The MANUAL light illuminates green when the system operates in the manual mode. The OFF SCHED DESCENT light (amber) illuminates when the aircraft descends before reaching the planned cruise altitude set in the FLT ALT indicator. An ALTN light illuminates green when the system operates in the alternate automatic mode. An amber AUTO FAIL light on the forward overhead pressurization panel provides an indication to the flight crew that a failure has occurred in the automatic pressurization control. The AUTO FAIL will come on if any of the following conditions occur:

- Loss of DC Power

– Controller fault in the operational controller

– Excessive rate of cabin pressure change (2 000 sea level ft/min)

– High cabin altitude (above 15 800 ft)

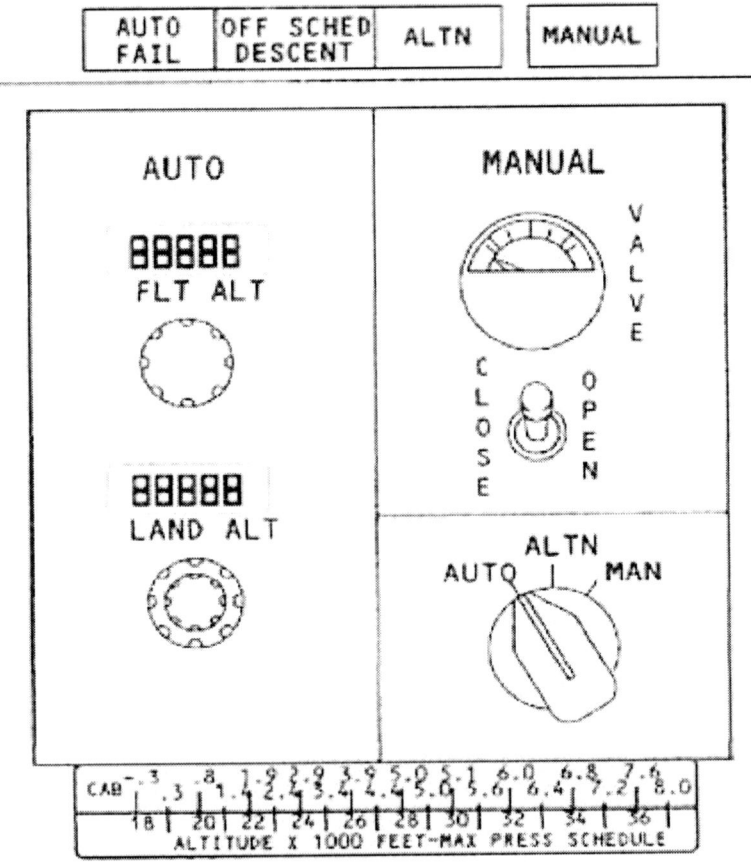

Diagram – Digital Pressure Control Panel (P5)

The system safety relief differential pressure is a maximum of 8.65 psi. The maximum allowable cabin pressure differential for takeoff and landing is 0.125 psi (236 ft below airport pressure altitude). The maximum cabin differential pressure (relief valves) is 8.65 psi.

Normal automatic operating differential pressure is: 7.8 + 0.1 psi above 28 000 ft pressure altitude, and 7.45 + 0.1 psi below 28 000 ft pressure altitude.

When, because of a malfunction during flight, the normal pressure differential cannot be controlled, the pressurization system may be allowed to operate for the remainder of the flight on the relief valves without affecting the safety of the airplane.

Outflow Valves

Two Outflow Valves (OFV), one forward and one aft, are installed in the aircraft which allow the excess air flow out of the cabin. If the OFVs are closed or nearly closed, the cabin pressure will increase. If the OFVs allow air to leave the cabin faster than it enters, the cabin pressure will decrease.

Pressurization Outflow

Cabin air outflow is controlled by the aft OFV valve, the forward OFV and the flow control valve (FCV). During pressurized flight, the flow control valve is closed, and the majority of the overboard exhaust is through the aft and forward OFV. A small amount is also exhausted through toilets and galley vents, miscellaneous fixed vents, and by seal leakages (Diagram – Digital Cabin Pressure Control System Schematic, Page 36).

Flow Control Valve

The flow control valve opens to exhaust the cooling air from the electrical and electronic (E & E) compartment overboard during ground operation, unpressurized flight and pressurized flight below a specified cabin differential pressure. When the FCV closes, air is directed around the forward cargo compartment liner for in-flight heating.

Outflow Valves (two units)

The aft OFV (main) is actuated by DC motors in any operating mode.

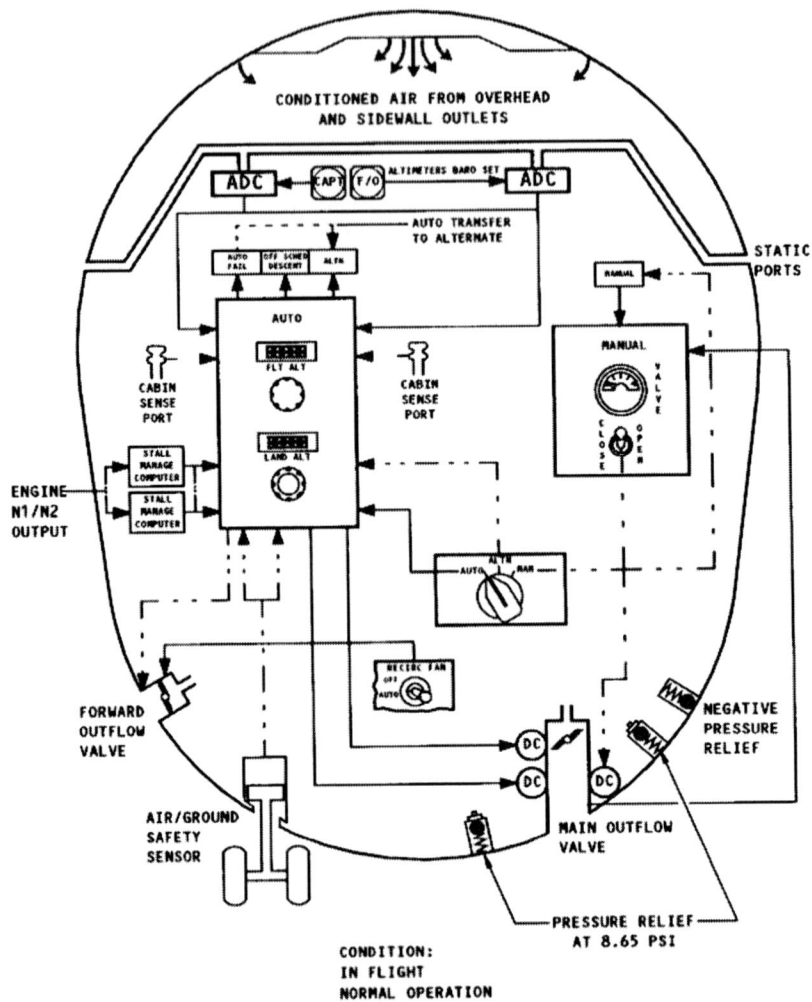

Diagram – Digital Pressure Control System Schematic

With either system, the forward OFV closes automatically to assist in maintaining cabin pressure when the aft OFV is less than 2 ± 1.5 degrees from being closed or the recirculation fan is operating. The forward OFV is the overboard discharge exit for air circulated around the forward cargo compartment. The aft OFV is the overboard exhaust exit for the majority of the air circulated through the passenger cabin. Passenger cabin air is drawn through foot level grills, down around the aft cargo compartment, where it provides heating, and is discharged through the aft OFV. The various positions of the aft OFV are as follows:

a) Fully open while airplane is on the ground unpressurized.

b) Partially closed while airplane is being pressurized on the ground

c) Modulating between open and closed for cruise, takeoff, and descent.

Equipment Cooling System

The equipment cooling system cools electronic equipment in the flight deck and the E & E bay. The system consists of a supply duct and an exhaust duct. Each duct houses a normal fan and an alternate fan. The supply duct brings cool air to the flight deck displays and to the electronic equipment in the E & E bay. The exhaust duct collects and discards warm air from the flight deck displays, the overhead and aft electronic panels, the circuit breaker panels in the flight deck and electronic equipment in the E & E bay.

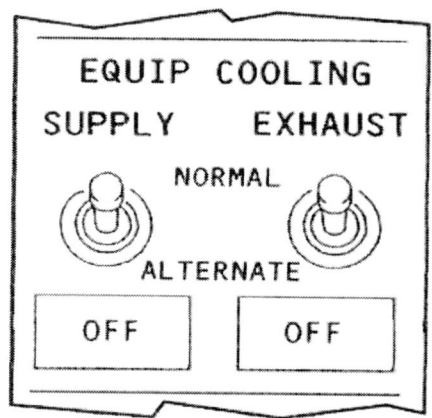

Diagram Equipment Cooling

Loss of airflow (mass flow) due to failure of an equipment cooling fan or low air density associated with excessive cabin altitude results in illumination of the related equipment cooling OFF light. Selecting the alternate fan should restore airflow and extinguish the OFF light within approximately 5 seconds, if the light is illuminated due to the failure of an equipment cooling fan. When either equipment cooling light is activated, the amber "MASTER CAUTION" and "OVERHEAD" indicator lights on the pilot's glareshield also illuminate. If an over temperature occurs on the ground, alerting to ground personnel is provided through the crew call horn in the nose wheel well.

On the ground, or with cabin differential pressure (Δp) less than 1.0 psi, the exhaust fan air is blown through a flow control valve and exhausted out at the bottom of the aircraft. With increasing airflow at greater cabin differential pressures, the flow control valve closes and exhaust air from the equipment cooling system is now diffused around the lining of the forward cargo compartment for in-flight heating of the forward cargo compartment. The recirculation fan system draws air from around the lining of the forward cargo compartment and air from the passenger cabin

Oxygen Systems

Two independent oxygen systems are provided, one for the flight crew and one for the passengers. Also portable oxygen cylinders are located throughout the aircraft passenger cabin for emergency use

The cockpit oxygen system uses quick-donning diluter demand masks/regulators located at each crew station. Oxygen is supplied by a single cylinder which is located in the forward cargo compartment. The cylinder has a maximum capacity of 115 cubic feet of free oxygen when pressurized to 1 850 psi. Oxygen pressure is displayed on the indicator located on the overhead panel.

The oxygen mask/regulator is stored in a box immediately adjacent to each crew station. To remove the mask from its stowed position the red levers around each mask must be squeezed with the thumb and forefinger. When squeezed, the harness inflates so it can be put easily over the face to allow breathing.

According to the FCOM (page 1.40.7, edition June 3, 2005) the observer's mask, regulator, and harness unit are the same as the flight crew's. However, during the investigation, it was determined that the Captain's and First Officer's oxygen masks were full face (integral goggles fitted) and the two observer masks were plain masks.

Oxygen flow is controlled by a regulator that is mounted on the oxygen mask. The regulator may be adjusted to supply 100% oxygen by pushing the NORMAL/100% selector.

The passenger oxygen is supplied by individual oxygen generators located at each Passenger Service Unit (PSU). A generator with two masks is located above each cabin attendant station.

The system is activated automatically when the cabin altitude climbs above 14 000 ft +/- 350 ft, or when the switch on the overhead panel is positioned to ON by the flight crew. When the system is activated the masks will drop, so the passenger has to pull the mask, put it on his/her face and start breathing. Once any of the four masks of a PSU is pulled, a continuous flow of oxygen begins. It lasts approximately 12 minutes and it can not be shut off. When the passenger oxygen system is activated, the following amber indicator lights illuminate on the flight deck: "MASTER CAUTION" and "OVERHEAD" on the pilot's glareshield and "PASS OXY ON" on the overhead panel

First aid and supplemental portable oxygen cylinders are installed at suitable locations in the passenger cabin. The number and location of portable oxygen cylinders varies with interior configuration. This aircraft was equipped with four cylinders located in the passenger cabin (two opposite the aft service door, and two next to the flight deck door). Each cylinder was fitted with a pressure gauge, pressure regulator and an On/Off valve.

The cylinders have a maximum capacity of 311 liters (11 cubic feet) of free oxygen when pressurized to 1 800 psi. The oxygen could be used either through a four liter per minute outlet, or through a two liter per minute outlet, resulting in an oxygen availability duration of 1h 17 minutes or 2h 35 minutes, respectively.

Cabin and Flight Deck Doors

There are four doors which provide access to the passenger cabin: two passenger/ crew entry doors on the left side of the fuselage (the forward entry door/1L and the aft entry door/2L) and two service doors on the right side of the fuselage (the forward galley service door!1R and the aft galley service door!2R). These doors are located approximately opposite the entry doors (See Diagram – Aircraft Doors Layout). They are similar in design and operation to the forward passenger/crew entry door, though slightly smaller. All four doors are inward-outward opening plug type doors and can be operated from inside or outside the airplane.

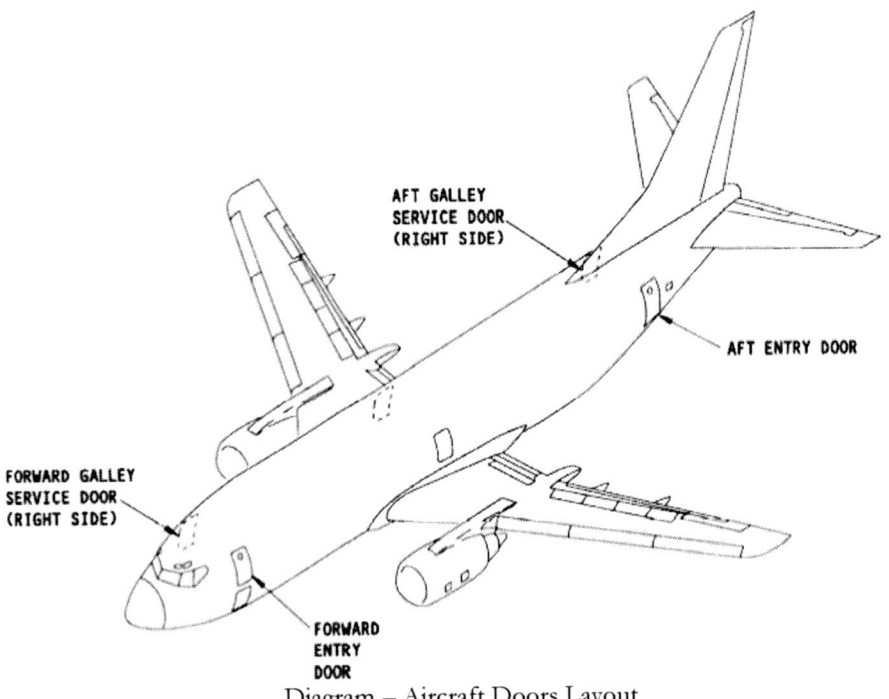

Diagram – Aircraft Doors Layout

An upper and lower hinge assembly supports the door on the forward edge of the door opening. The door is opened by manually operating the centrally located handle, to cause an internal mechanism within the door to release roller latches on the door from latch fittings on the door jambs, fold the gates on the top and bottom inward, and move the door to the most inward position. The door is then manually swung through the door opening and stowed in the open position by the engagement of the hold-open lock mechanism attached to the hinge torque tube. (Diagram - Passenger Entry/Galley Services Door).

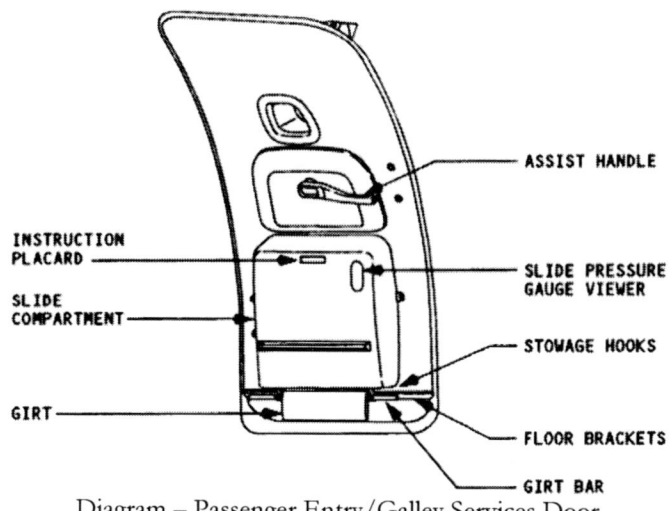

INSTRUCTION PLACARD

SLIDE COMPARTMENT

GIRT

ASSIST HANDLE

SLIDE PRESSURE GAUGE VIEWER

STOWAGE HOOKS

FLOOR BRACKETS

GIRT BAR

Diagram – Passenger Entry/Galley Services Door

The door will close by applying the opening procedure in reverse, after the hold-open lock mechanism has been disengaged. When the door is in the closed position and the passenger cabin is pressurized, it will be "plugged" in the door opening, firmly secured by the latching elements, not allowing the pressurized air to escape, provided that all the surrounding sealing material is intact.

A door warning system proximity switch is attached to the door and door frame alerting the flight crew in case a door is not properly closed and latched. Assist handles are provided to give additional control during the door operating cycle.

Flight Deck Door

This door isolates the flight compartment from the passenger compartment and meets requirements for ballistic penetration and intruder entrance. Features of the door (such as removable panels, etc.) are designed to comply with or to aid compliance with decompression or egress requirements. Components of the flight deck access system include an emergency access panel, chime module, three-position door lock selector, two indicator lights, and an access system guarded switch. The emergency access panel includes a six button keypad for entering a numeric access code along with red, amber and green lights.

Meteorological Information

The weather (METAR) at Larnaca Airport on 14 August 2005 at 06:00 h was: wind 350 degrees 2 kt, CAVOK (ceiling and visibility OK), temperature 32 °C, dew point 21 °C, QNH 1008 hPa, and NOSIG (no significant change).

The en route weather from Larnaca Airport via EVENO to RIPLI was good. There was no significant weather, the wind was 250 – 270 degrees (southwest-west) at 37 – 50 kt and the temperature was from -38 °C to -40 °C. From RIPLI to KEA VOR the wind was southwesterly at 40–55 kt and the temperature was from -39 °C to -42 °C.

The weather forecast for Athens International Airport on 14 August 2005 for 04:00 – 13:00 h was: wind variable at 5 kt, becoming (08:00 – 10:00 hours) wind from 020 degrees at 12 kt. The METAR at 06:50 h showed wind from 140 degrees at 6 kt, CAVOK (no clouds and unlimited visibility), temperature +32 °C, dew point +13 °C, QNH 1009 hPa, and NOSIG (no significant change).

Aids to Navigation

On departure from Larnaca, flight HCY522 used Larnaca VOR (112.80 MHz) and then the en route navigational aid, Rhodos (RDS) VOR (115.80 MHz). The navigational aid on airway UL995 was KEA VOR (115.00 MHz). At Athens International Airport, the aircraft overflew runway 03L at FL340 and proceeded back to KEA VOR where it joined the holding pattern. The navigational aids were functioning normally and no discrepancies were reported before or after the accident.

Detailed information on navigational aids and landing aids at Athens International Airport was available in Aeronautical Information Publication Greece (AIP-Greece). The navigational aids and landing aids at Athens International Airport were functioning normally.

Communications

Flight HCY522 communicated with Larnaca Tower in English with some phrases in Greek on frequency 121.2 MHz, and with Nicosia ACC also in English on frequency 125.5 MHz. The last communication with Nicosia

ACC was at 06:11:45 h. These communications were normal and were recorded in the ATS facilities in Cyprus. The communications were transcribed from the recordings.

Athinai ACC called the flight on frequency 124.475 MHz and on the emergency frequency 121.5 MHz but there was no response. The leader of the F-16 fighters also called the flight on the emergency frequencies 121.5 MHz and 243 MHz but received no response.

Flight HCY522 also communicated with the company Dispatcher office on the company frequency, 131.2 MHz. The communications with the Dispatcher Office were not recorded. There was no requirement by Cyprus DCA, JARs or ICAO Annexes for the recording of company communications with airborne aircraft.

Radio Communication Failure

After six unsuccessful calls on frequency 125,5 MHz, Nicosia ACC called flight HCY 522, per standard procedures, to "squawk STAND–BY", and then called on the emergency frequency 121.5 MHz to rule out the possibility of one-way radio failure (aircraft receiving but unable to transmit). When the efforts to communicate with the aircraft proved unsuccessful, according to the Nicosia Planner he called, by telephone, the Planner Controller at Athinai ACC and informed him that he had no contact with the flight.

According to provisions of ICAO Doc 4444, Air Traffic Management, regarding radio communications failure (RCF):

Paragraph 11.4.1.3.1 – *When an air traffic services unit is aware that an aircraft in its area is experiencing radio communication failure, an RCF message shall be transmitted to all subsequent ATS units along the route of flight which have already received basic flight plan data (FPL or RPL) and to the aerodrome control tower at the destination aerodrome, if basic flight plan data has been previously sent.*

Paragraph 11.4.1.3.2 – *If the next ATS unit has not yet received basic flight plan data because it would receive a current flight plan message in the coordination process, then an RCF message and a current flight plan (CPL) message shall be transmitted to this ATS unit. In turn, this ATS unit shall transmit an RCF message and a CPL message to the next A TS unit. The above process shall be repeated progressively from centre to centre up to the first ATS unit along the remaining route of flight to which basic flight plan data has already been sent.*

Paragraph 15.3.7 – *As soon as it is known that an aircraft which is operating in its area of responsibility is experiencing an apparent radio communication failure, an air traffic services unit shall forward information concerning the radio communication failure to all air traffic services units concerned along the route of flight. The ACC in whose area the destination aerodrome is located shall take steps to obtain information on the alternate aerodrome(s) and other relevant information specified in the filed flight plan, if such information is not available.*

Handover from Nicosia ACC to Athinai ACC

Air Traffic Control handover from Nicosia ACC to Athinai ACC was performed "procedurally" due to incompatibility of the systems used by Athinai and Nicosia ACCs, which precluded automatic handover.

An "ESTIMATE" message was sent, initially through AFTN and then by oral confirmation by telephone in the form of a "VERBAL ESTIMATE" (call sign, SSR code, estimated time over the boundary of the FIRs involved and cleared flight level), in order to confirm the data received earlier through the AFTN message.

According to the Letter of Agreement (LoA), regarding coordination procedures between the two ACCs, at entry point EVENO, on the boundary of the two FIRs, the handover of control and communications of any flight from Nicosia FIR was performed at even flight levels with 20 NM radar separation using airway UM60 1.

According to AIP-Greece, any aircraft entering the Athinai FIR must report to Athinai ACC. If that did not occur, according to regulations, Athinai ACC should make radio contact with the aircraft. Because the aircraft did not report its entry, and because Athinai ACC did not call the flight either, the transfer of communication of the flight in question from Nicosia ACC to Athinai ACC was not complete.

Airport Information

Larnaca Airport had one runway 04/22. Its dimensions were 45 x 3.050 m, the surface was asphalt and the elevation was 8 ft (above MSL). The provision of Approach Control Service within the Larnaca Control Zone was provided by Larnaca Tower, simultaneously with Aerodrome Control Service on frequency 121.2 MHz.

Air traffic control radar service within Nicosia FIR was provided by three control sectors: West Sector, East Sector, and South Sector. There are two working positions in the West Sector which handled flight HCY522; Executive (Radar Controller) and Planner (Procedural Controller).

Athens International Airport "El Venizelos"

Flight HCY522 intended to land at Athens International Airport. The airport had two parallel runways. The dimensions of runway 03L/21R were 45 x 3.800 m and the dimensions of 03R/21L were 45 x 4.000 m. Both runway surfaces were asphalt. The elevation was 94 ft (above MSL).

Air traffic control radar service within Athinai FIR was provided by two Area Control Centers (Athinai ACC and Makedonia ACC) using an automated radar system (PALLAS). The functions "assume," "accept," "handover," "handout" in the radar control procedures were carried out by the Radar Controller at the FIR boundaries.

Athinai ACC comprised ten control sectors. Two control sectors, namely ACC_2 LOW and ACC_9 UPPER, provided ATC radar service to the southeast part of Athinai FIR which handled flight HCY522. Depending on the traffic load, the two control sectors could be combined into a single control sector, ACC_2, which provided ATC radar service on 124.475 MHz.

There were six shifts to cover the 24-hour watch at Athinai and Makedonia ACCs. In each shift there was a Supervisor. There were three working positions in each control sector, i.e. the Executive (Radar Controller), the Planner (Procedural Controller) and the Assistant Controller. Depending on the traffic and available personnel, the Assistant Controller position may not be staffed and the duties handled by the Planner controller.

Flight Recorders

The aircraft was equipped with a Fairchild Model A100S CVR Part No. S100-080-00, Serial No. 01768. The CVR had four independent channels to record a minimum of 30 minutes of audio, including the Captain, First Officer and cockpit area sounds. The CVR was severely damaged by impact and the memory module had separated from its crash/fire protection case.

A portion of the CVR outer case was found on 15 August 2005 near the left wing tip; the remainder of the outer case was found the next day. After an intense search, using a photograph of the memory module to facilitate the recognition of the module in the search effort, the module was found on 19 August 2005.

The CVR was taken to be transcribed at BEA (Bureau d' Enquétes et d' Analyses) in France. The Flash Card Survivable Store Unit (FCSSU) was opened and the memory board was extracted. The memory board was connected to the acquisition part of the Solid State CVR by a cable from the FCSSU (memory-protected module). The cable was found in good condition; a new connector was plugged into the cable, which was then connected to BEA'S A100S unit, which was used as a reader. After the data extraction was completed, the last 30 minutes of the recording was obtained.

Front cover of the outer case
of the CVR

Memory module of the CVR

The FDR was an Allied Signal model 980-4700-003, Serial No. 1165. It was a solid state memory FDR with 50 hours of recording capacity (when used at the rate of 64 words per second) and recorded 254 parameters.

The FDR was found near the right aft service door and was recovered from the accident site on 14 August 2005, the day of the accident. It was taken to the readout facilities of BEA and the read out began on 17 August 2005. The recorder was in good condition, with no evidence of impact damage. However, the outside casing, including the outside of the acquisition unit, was blackened by fire expose and the paint was melted. Since it bore signs of fire exposure, the protected unit was removed from the FDR and connected to a new acquisition unit. It was then connected to the readout equipment, in order to download the raw data file. The data file was synchronized using BEA software.

CHAPTER 2

WRECKAGE AND IMPACT INFORMATION

After on-site examination, documentation, and photographing, all parts and components retrieved from the accident site were relocated to a guarded storage facility at the Athens airport where they were properly labeled.

Distribution

The wreckage was distributed generally from west to east, over two rocky hills for a straight-line distance of about 800 m (Diagram - Wreckage Distribution Map). The first component found was a drain mast where the fuselage first impacted the ground. Tufts of fiberglass thermal-acoustic insulation batting were in the immediate area.

Two adjacent dirt paths (gouges) were found at the initial impact area, which matched the spacing of the fuselage to the right engine. The dirt paths passed through low vegetation on the first uphill slope. The wider path was found to be consistent with the 3.7 m width of the fuselage and the adjacent path to the right was consistent with the 2.4 m width of the engine. A third dirt path was found to the left, 10 m down-track from the initial contact of the fuselage and right engine. The location of this dirt path was consistent with the spacing of the left engine from the initial impact of the fuselage and right engine. A 35 cm wide slice through the vegetation began 9 m to the right of the fuselage centerline and 5 m down-track from the initial contact of the fuselage and right engine.

Initial impact area and first two contact paths on first uphill slope. The path with the width of the fuselage begins in the center of the photo and the path with the spacing and width of the right engine begins near the right side.

The initial points of contact were on an uphill slope that rolled off to the left. The right engine track turned to the left and crossed the fuselage path. After the initial and flatter portion of the slope was a slight upward step, followed by the path of the left engine passing over untouched vegetation. This was followed by an in-line impact in the path of the left engine and concentration of fuselage debris in the fuselage path, where the slope turned upward. The fuselage path turned about 2.5 degrees to the left in this area. The area included the upper forward corner of the aft cargo door and the top of the nose landing gear well. The clear lens from the left wing tip was found to the left of the path and further up the hill, about 2 m off the left wing tip on the right side of the path. A secondary path from the upward step led to the nearly complete tail assembly, which was found to the right of the main wreckage path.

The tail assembly near the crest of the hill.

The 470 m elevation at the crest of the first hill had a field, which rolled off to the left of the path of the debris. A generator and constant speed drive assembly were found toward the eastern portion of the field and nearby was a fragment of aft pressure bulkhead, nose landing gear structure containing the trunnion mount, and a cargo compartment liner. The crest of the first hill was followed by a downward slope that then dropped into a ravine (see photo below). Sequentially down the slope were an air conditioning pack heat exchanger, with a hand-written "R/H FWD," the negative pressurization valve, the first observed complete thrust reverser actuator (stowed), the aft cargo compartment floor, the left and right fuselage skins with most of the roof attached to the right window belt, and a door slide.

Viewed backward from the second hill, this photo looks at the downward slope of the first hill into a ravine. The tail assembly is at the extreme upper left corner, beneath the two trees. The skin of the right aft fuselage and top of fuselage is the white material at the right edge of the photo, above the green vegetation. The upper edge of the first ravine is visible along the bottom of the photo, which then drops out of sight.

The ravine dropped to 380 m above sea level. The ravine had side channels and debris was found collected at the bottom, such as the bottom third of the aft cargo door, fragments of the main landing gear compartment, and parts of the floor of the forward cargo compartment. Similarly, an engine fuel control and pump were found in the ravine beneath an engine generator, thrust reverser frame, and gearbox fragments on the uphill slope to the second crest.

Down-track from the first ravine was a nearly vertical rise up to a slope that crested at the top of the second hill, where the cockpit and avionics compartment were found at about 430 m above sea level (See photo below).

The downhill slope from the first peak is to the left and the upward slope to the peak of the second hill is to the right. The dark debris in the center of the upward slope to the right is half of the left engine.

On the uphill to the second ridge, was a main landing gear strut with a diagonal scar on the up-lock. The ridge curved and beneath the strut in the bottom of the ravine was a set of main landing gear tires. A half of the left engine core was found near an engine pylon that contained a closed pressure regulating shutoff valve (PRSOV) that was missing the housing for the actuation diaphragm.

The right wing tip is at the lower left and the left wing isat the right edge of the photo. The center section and main gearwell is at top right.

Uphill and south of the engine half and pylon was a second pylon, missing a PRSOV component that was similar to the one found attached to the right engine. Most of the PRSOV was with the pylon and was in the closed position. Near the uphill pylon was an engine fan ring. The left engine half found later at the bottom of the second ravine had a fan ring nearby. Near the crest of the second hill were most of the avionics compartment, cockpit, air stair door, the two forward cabin doors, and the forward cargo door. Past the ridge of the second hill, were the auxiliary power unit (APU) motor and firewall, the aft cabin doors, and the nearly complete right engine core.

The nearly complete wings assembly was found about halfway down the second slope. The main landing gear well was burned out and was found about 12 m uphill from the wings. The wings were found folded so that the tips were downhill. Following a deep impact into the downhill slope, a series of engine components led downhill to half of the left engine.

Aircraft System Components

The aircraft's cockpit instruments, primary and secondary flight control systems and environmental/pressurization control system components were identified, photographed and documented.

Overhead Cockpit P5 Panel

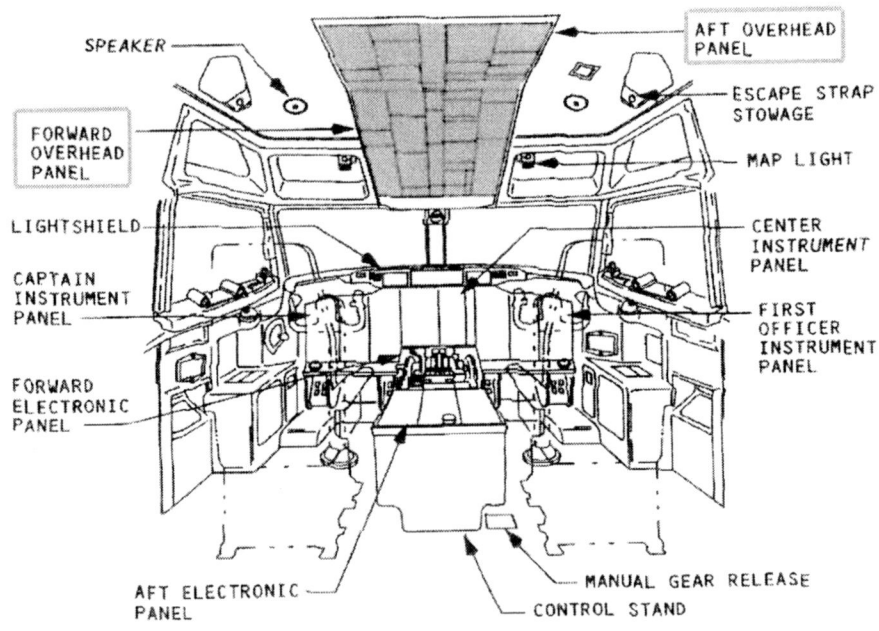

Position of overhead P5 panel

The forward overhead P5 panel was located in the area of the cockpit within the wreckage at the accident site. The panel was accessed by removing the surrounding wreckage. Once the panel was located, low air pressure was used to remove the accumulated dirt, dust and light debris in order to visually examine the controls.

P5 Flight Deck Control Panel - toggle switches, selectors, switches positions:

Control	Position
ALTERNATE FLAPS	OFF
YAW DAMPER	OFF
VHF NAV	Positioned to NORMAL
IRS	Positioned to NORMAL
EFI	Positioned to NORMAL
FUEL PUMPS	Both left and right switch positioned UP Number 1 ON, and Number 2 ON
DC switch	Positioned to BAT
GALLEY	Switch positioned to OFF
DRIVE TEMP	Positioned IN
GRD PWR	Switch positioned to OFF
APU GEN	Switch positioned to GEN 2 GEN 1 Switch positioned to OFF LH APU GEN Switch positioned to OFF RH APU GEN Switch positioned to OFF GEN 2 Switch positioned to OFF
EQUIP COOLING	Supply Switch positioned to NORMAL Exhaust Switch positioned to NORMAL
NO SMOKING	Switch positioned to AUTO
FASTEN BELTS	Switch positioned to AUTO
WIPER	Switch positioned to OFF
WINDOW HEAT	Left side switch positioned to ON Right side switch positioned to ON
OVHT/PWR TEST	Switch positioned to neutral
PITOT STATIC HEAT	A switch positioned to OFF B switch positioned to ON
WING ANTI-ICE	Switch positioned to OFF
ENG ANTI-ICE	Both number 1 & 2 switches positioned to OFF
HYD PUMPS	A ENG 1 ELEC 2 Both positioned to ON B ENG 2 ELEC 1 Both positioned to ON

The entire P5 panel was photographed and the position of the controls (toggle switches, selectors, switches, and indicators) were visually examined and documented. The controls were not activated.

At the storage facility, the surface of several control panel faceplates were carefully wiped down with a damp cloth to more accurately read the nomenclature for each control. No controls were activated at the storage facility. The table on page 56 provides the documentation of the controls as found at the accident site and the storage facility.

P5 Air Conditioning Panel

The P5-10 air conditioning panel remained in the P5 overhead panel. The panel was photographed and the positions of each toggle switch and the duct pressure indicator were documented.

Air Conditioning Panel

The left engine bleed toggle switch (BLEED 1) was found in the OFF position. The right engine bleed toggle switch (BLEED 2) was visually found in the OFF position. The APU toggle switch was found in the OFF Position. The isolation valve toggle switch was found in the AUTO position. The left air conditioning pack switch appeared to be in the AUTO position. The position of the right pack switch could not be determined because of impact damage. The air conditioning panel was packaged and shipped to the Boeing Company, along with other systems components, for additional examinations.

Cabin Pressurization Panel

The P5-5 cabin pressurization panel was found intact and remained in the P-5 overhead panel. The position indicator of the cabin pressure mode selector was found rotated past the MAN (manual) position (on the faceplate) by approximately 7.0 mm. It was later determined by the Boeing EQA that the damage was consistent with physical impact.

Pressurization Panel

The selector had an abrasion, in line with the selector line, on its right side at the lower corner. The panel face plate in the vicinity of the AUTO nomenclature had a 180 degree circular impression of the selector base. The center mark on the selector was about 10 degrees from the start of the circular impression

The cabin pressurization panel was packaged and shipped to the Boeing Company, along with other systems components, for additional examinations.

Cabin Pressure Controllers

The aircraft was equipped with two cabin pressure controllers manufacturer by Nord-Micro. One controller was severely burned with its identification plate missing. The other controller had slight heat damage. Both cabin pressure controllers were removed from the accident site and transported to Nord-Micro in Frankfurt, Germany for a Non-Volatile Memory (NVM) download.

Forward Outflow Valve

Forward Outflow Valve

The Forward Outflow Valve, manufactured by Grimes Aerospace Co., was found intact and connected to its ducting. The manufacturer Commercial and Government Entity (CAGE) code number on the actuator was 72914, which indicated that the actuator was made by Honeywell (the parent company of Grimes Aerospace Co.). The part number and serial number of the actuator were 2535-3 and MTV0580, respectively. The actuator was a subassembly of the Forward Outflow Valve, P/N 32- 2684-007. The valve position indicator lever remained in the closed position, and an electrical connector remained connected to the valve assembly. A tag on the valve stated *"installation eligibility Boeing 74 7-200, 74 7-300, 747-400."* According to Boeing, this valve is the *"approved 737 forward outflow valve as verified by part number 32-2684- 007 ... the part is also approved for use on the 747."* The butterfly valve contained within the assembly was in the closed position.

Aft Outflow Valve

The actuator/motor assembly and both the forward and aft door gates of the aft outflow valve (AOV) were found at the accident site. The actuator/motor assembly was separated from each of the door gates.

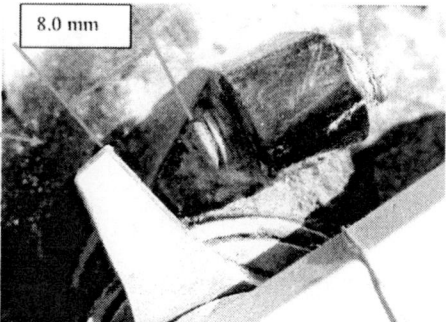

Aft Outflow Valve at the accident site

The condition of the actuator/motor assembly was documented and photographed at the accident site. The control arm and position indicator remained attached to the motor assembly. Using hand pressure, a small force was applied to the control arm; it was physically locked in place and not able to rotate freely. At the storage facility, a measurement was taken between the closed position stop and the control arm position lever; the distance between the two was 8.0 mm.

Pneumatic System Components

The aircraft pneumatic system components, which were found in various locations at the accident site, were examined and documented. No pre-accident damage or malfunctions were noted during these examinations.

Flight Deck Oxygen Cylinder

The flight deck oxygen cylinder was found heat damaged and the regulator was missing.. The shutoff valve assembly was observed to be loose on the cylinder and it could easily be rotated in the clockwise and counter clockwise direction. At the storage facility, the cylinder was visually examined and the regulator valve was physically rotated. Using hand pressure, a force was applied to the regulator valve in the counter-clockwise (valve open) direction; the valve did not rotate. Using hand pressure, a force was applied to the regulator valve in the clockwise (valve closed) direction; the valve rotated six revolutions. The cylinder was empty. The last hydrostatic check had been carried out in September 2003 (09/03 dye stamp on cylinder neck).

Flight Deck Oxygen cylinder

Flight Crew Oxygen Masks

One flight crew oxygen mask storage box was found at the accident site in the area of the flight deck. The mask storage box was found crushed; the mask/regulator was not in the box. Two flight deck plain oxygen masks were also found in the vicinity of the flight deck. Both oxygen masks were observed to not have integrated goggles; the face mask would only cover the mouth and nose. Neither mask was removed from its plastic bag (in which it was placed upon recovery from the site) during the documentation procedure, as the masks were then sent for DNA testing and thereafter to Zodiac-Intertechnique for functional testing.

Portable Oxygen Bottles

A total of four portable oxygen bottles were located at the accident site. In order to determine if the regulator was in the open or closed position, the initial position of the valve handle was marked at the storage facility and then the valve handle was initially rotated clockwise (closed position). If the valve handle did not rotate clockwise, the Board documented that the regulator was in the closed position and no further action was taken. If the valve handle rotated clockwise, the Board documented the number of revolutions the valve handle rotated. The following describes the condition of each of the four bottles:

- Bottle number 1: The bottle had been exposed to heat. Paint was missing and the exterior of the bottle appeared to be slightly rusted. The shut-off valve was rotated one revolution clockwise (360 degrees) until it was in the fully closed position.

- Bottle number 2: The bottle assembly was in its normal appearance with green paint on its exterior surface and the shut-off valve colored yellow. The shut-off valve was rotated one revolution clockwise (360 degrees) until it was in the fully closed position.

- Bottle number 3: The bottle had been exposed to heat. The paint was missing and the exterior of the bottle appeared to be slightly rusted. The shut-off valve was found in the fully closed position.

- Bottle number 4: The bottle had been exposed to heat. The paint was missing and the exterior of the bottle appeared to be slightly rusted. The shut-off valve was rotated 1 1/2 revolutions clockwise to the fully closed position.

Portable Oxygen Bottles

Engines

Following the thrust reverser and gearbox fragments up the first slope, a third of an engine inlet was found where the left edge of the wreckage trail reached the top of the first ridge and about 46 m further on the same ridge was another two-thirds of the engine inlet.

The left engine was primarily found in two sections, on the uphill slope to the second ridge and at the bottom of the second ravine. The broken end of the shaft was visible and had broken at an approximately 45-degree angle to its surface. The right engine was nearly complete, slightly down-track from the crest of the second ridge.

Neither engine had an outward penetration of the compressor or turbine sections. The visible compressor and turbine blade tips did not have evidence of rotational scratches at the tips. Most blades were straight, with some bending forward mid-span, although some tips of the right engine were slightly bent aft. No melted metal splatter was found in the exhaust areas of either engine or the APU. The output shaft of the APU had a bend of about 10 – 20 degrees.

Thrust reverser actuators were found along the path of debris prior to the slope leading down to the second ravine. All of the thrust reverser actuators were found in the retracted positions.

Landing Gear

The nose landing gear was found in two primary sections, with the actuator on the first upward slope and the strut near the crest of the second hill. One wheel and tire were found at the bottom of the ravine. The actuator was found extended. The structure of the nose landing gear box was found in the area of the actuator on the first upward slope.

The main landing gear struts were found near the top of the second crest on the uphill slope. The strut to the north was missing the pair of wheels and brakes, which was subsequently found down-slope in the first ravine. The uplock roller on the northern strut had a diagonal scar. The strut to the south was found in fragments and the uplock roller was not identified.

The main landing gear well had experienced extensive post-crash fire damage and the uplock was not identified.

Doors

All four cabin doors (1L, 2L, 1R and 2R) were recovered from the accident site, having suffered severe impact damage.

A detailed examination of each aircraft door was conducted at the accident site and revealed no evidence of pre-impact defects or mechanical failures related to the accident circumstances.

The flight deck door was recovered from the accident site separated from its fuselage structure and with some damage to its lower portion. No pre-impact damage was noted on either side of flight deck door faces or lower blow out panel. The door upper blow-out panel was missing. The flight deck access system guarded switch on the door side post fragment was observed to be positioned to "NORM". The flight deck door deadbolt was found in the "UNLOCKED" POSITION (vertical).

CHAPTER 3

MEDICAL AND PATHOLOGICAL INFORMATION

Medical Information

The remains of 118 of a total of 121 aircraft occupants (115 passengers and 6 crewmembers) were recovered and examined by the Greek Forensic Medical Services. According to the Forensic Pathologist, the remaining three bodies were probably consumed by the post-accident fire. Sixty-two bodies were severely burnt and were identified using dental records and DNA tests. According to the Pathologist's report, the cause of death for all on board was determined to be the multiple injuries due to impact, in addition to the extensive burns for 62 of them.

All passenger and cabin crew samples tested negative for carbon monoxide and volatile poisons. The forensic report concluded that the aircraft occupants had heart function during the impact. The report noted that this did not necessarily imply that they were alert. The report further estimated that they were in deep non-reversible coma due to their prolonged exposure (over 2.5 h) to the high hypoxic environment.

Toxicological samples were taken from all six crewmembers and 108 passengers. The Captain's samples (obtained on 18 August 2005) tested negative for major drugs of abuse, volatile poisons, and prescription and over-the-counter medications. Due to the presence of extensive burns, the determination of blood alcohol level was not possible.

The Captain's heart muscle samples revealed the presence of minor atherosclerosis (40% obstruction) compatible with his age. A histological examination revealed the presence of recent myocardial ischaemia.[1])

The First Officer's samples (obtained on 15 August 2005) tested negative for carbon monoxide, volatile poisons, major drugs of abuse, and prescription and over-the-counter medications. Although ethanol was detected (34 mg/dl, or 0.034 % weight/volume - also known as blood alcohol content), the toxicological reports stated that *"the time period between the death and the collection and the analysis of specimens (24 hours) may have resulted in post-mortem ethanol production."* The First Officer's heart muscle samples revealed the presence of extensive atherosclerosis (90% obstruction in the anterior descendant and circumflex coronary artery). A histological examination revealed the presence of recent myocardial ischemia.

The Department of Cardiology of the Hellenic Air Force Medical Centre predicated that *"On the basis of the data that were given to us, such as the height of the flight, the fact of the existing heart function (pump function) upon crashing, and the fact that there is a similar pathologoanatomical image both in the 'suffering' heart (myocardium) of the co-pilot, and in the 'healthy heart' of the pilot, we estimate that the brain hypoxia was the dominant and determinant cause that incapacitated the flying crew, with the findings of the heart being the matter of course and epiphenomenon of the prolonged hypoxia."*

According to his medical records and statements by his next of kin and friends, the First Officer was asymptomatic and had a normal lifestyle. The examinations performed during his regular aeromedical examination were within the normal values. The ECGs were evaluated by a UK CAA cardiologist as normal. He had no risk factors for cardiovascular disease (family history, dislipidaimias, diabetes, hypertension, etc.). The aeromedical examination was performed per JAR-FCL 3 procedures.

The cardiologist estimated that by-pass circulation had developed in the First Officer's cardiovascular system, thus providing satisfactory heart function. This type of circulation was a functional and not anatomical phenomenon, and may not be detected during the forensic examination.

Physiological and Psychological Effects of Low Cabin Pressure

An extensive summary of basic aeromedical facts is given in the *Aeromedical Handbook for Aircrew* (Dobie, 1972). (Further information

64

can be found in comprehensive textbooks of Aviation Medicine, e.g. Armstrong 1961, Gillies, 1965, DeHart, 1985, Ernsting, 1999). A recent compilation of both physiological and psychological effects of low pressure is given in the altitude chapter of *Exercise Physiology* by McArdle, Katch & Katch, 2001.

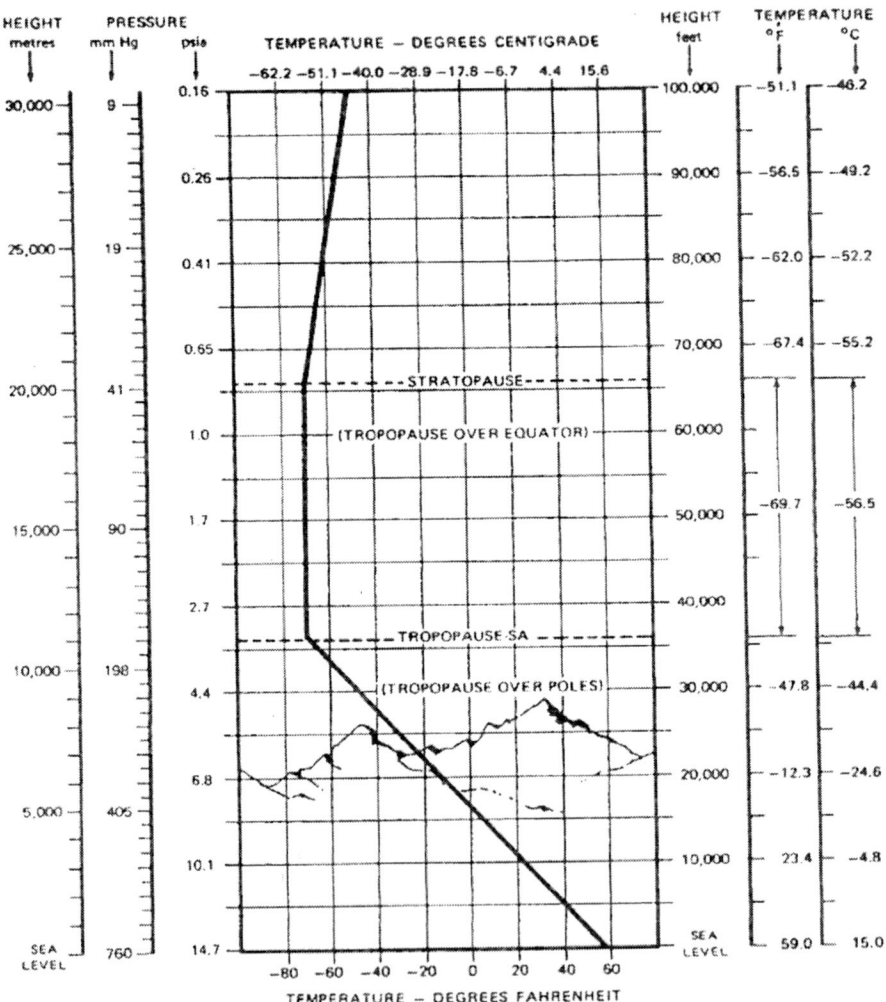

A diagrammatic representation of the properties of the standard atmosphere showing the variation in the height of the tropopause at different latitudes

With increasing altitude and since oxygen concentration is constant, pressure and temperature of the atmosphere decrease as does the oxygen partial pressure, which is essential for sufficient oxygenation of all higher life forms (See Standard Atmosphere Chart, page 64).

Hypoxia

The amount and pressure of oxygen delivered to the tissues is determined by arterial oxygen saturation, by the total oxygen-carrying capacity, and by the rate of delivery to the tissues.

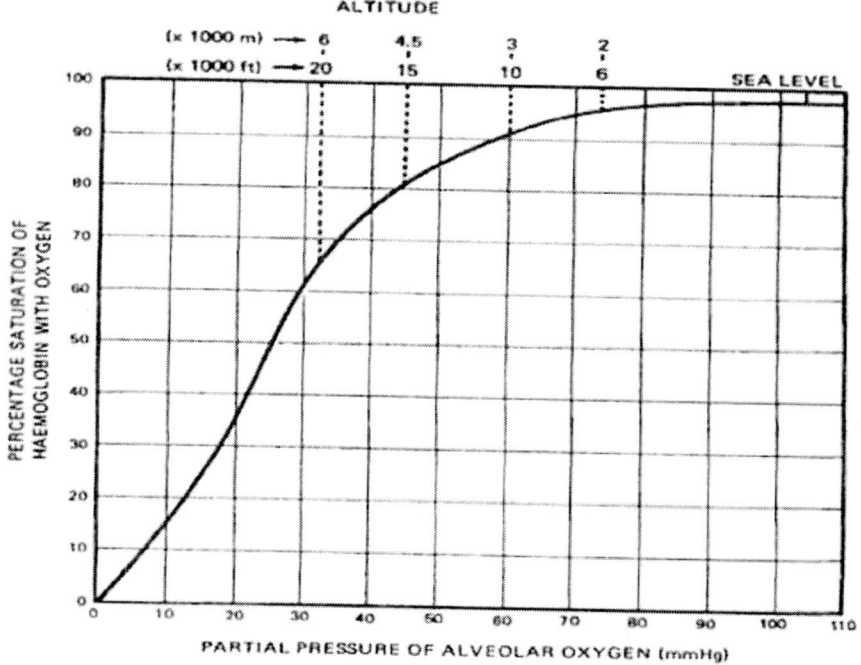

Diagram Arterial Hemoglobin Oxygen Saturation at Altitude

The oxygenation of an organism takes place in the lungs; at a sea level pressure of 760 mm Hg, the oxygen concentration of approximately 21% in the air leads to an oxygen partial pressure of 210 hPa in dry air, sufficient for proper oxygenation in all organisms adapted to these conditions. However, this is not the partial pressure effective in the lungs, since, in the process of breathing the air is warmed up to the body core temperature of 37°C and saturated with water to a relative humidity of 100% (approximately 63 hPa of water vapour), thus decreasing the oxygen contents. Furthermore, a certain amount of oxygen is exchanged with carbon dioxide, leading to an effective oxygen partial pressure of approximately 140 hPa in the lung alveoles. This alveolar oxygen partial pressure (p_AO_2) is the driving force for the blood oxygen saturation of haemoglobin (SaO_2), which is usually 96-98%.

When ambient pressure decreases, also p_AO_2 decreases, leading to a lower haemoglobin oxygen percentage of saturation and finally hypoxia. However, with the onset of hypoxia a number of adaptation processes start, leading to a higher p_AO_2 than expected with normal breathing. The Diagram of Arterial Hemoglobin Oxygen Saturation at Altitude (see page 68) shows the decrease of arterial haemoglobin oxygen percentage of saturation both with alveolar oxygen partial pressure and altitude. Apparently, SaO_2 does not change very much up to about 5 000 ft; at 8.000 ft it is approximately 4-5% lower than at sea level.

The decrease in SaO_2 leads to both physical and mental performance degradation. In aircraft operations, where muscle work is not the major requirement, the decrements in mental capabilities are most important for the flight crew in relation to safe aircraft guidance. The diagram of Haemoglobin Oxygen Saturation and Performance shows a collection of corresponding findings.

Hypoxia, defined as an insufficient supply of oxygen, can result from any one of these factors. The following classic types of hypoxia have been distinguished:

1.Hypoxic hypoxia results from an inadequate oxygenation of the arterial blood and is caused by reduced oxygen partial pressure.

2.Anemic hypoxia results from the reduced oxygen-carrying capacity of the blood, which may be due to blood loss, any of the anemias, carbon monoxide poisoning, or by drugs causing methemoglobinemia.

3.Stagnant hypoxia is caused by a circulatory malfunction which results, for example, from the venous pooling encountered during acceleration maneuvers.

4.Histotoxic hypoxia_results from an inability of the cells to utilize the oxygen provided when the normal oxidation processes have been poisoned such as by cyanide. There is no oxygen lack in the tissues, but rather an inability to use available oxygen, with the result that the P_AO_2 in the tissues may be higher than normal. Therefore, it is not true hypoxia by the definition used here.

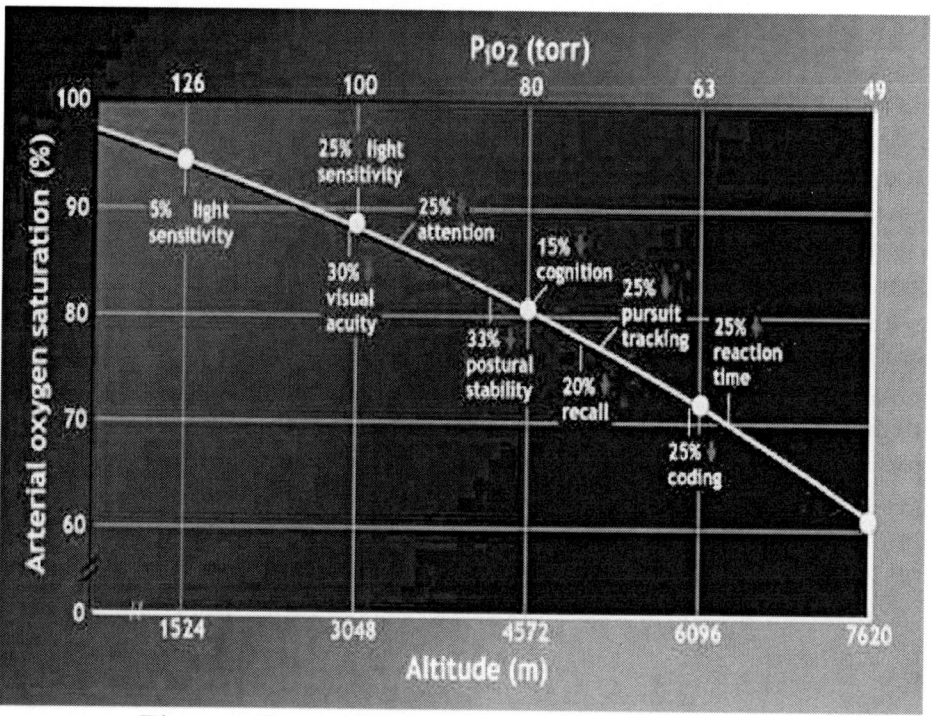

Diagram – Haemoglobin Oxygen Saturation and Performance

The most common type of hypoxia encountered in aviation is *hypoxic hypoxia*. This results from the reduced oxygen partial pressure in the inspired air caused by the decrease in barometric pressure. Other types may also affect air crewmen, such as anemic hypoxia as seen in carbon monoxide poisoning and stagnant hypoxia resulting during various acceleration profiles.

Symptomatology

Exposure to high altitude with consequent hypobaric hypoxia and reduced oxygen supply to the central nervous system causes a variety of neuropsychological symptoms (Koller et al., 1991; Lieberman et al., 1994). The most commonly experienced symptoms recognized as hypoxia are cognitive impairments (Blanchet et al., 1997; Bonnon et al., 1995; Cable, 2003; Cahoon, 1972; Macintosh et al., 1988; Shepard, 1956; Shukitt-Hale et al., 1998). In addition to deterioration of cognitive performance, emotional changes including euphoria, irritability, hostility, and overconfidence have been reported (Shukitt-Hale & Banderet, 1988). Exposure to altitudes above 19.700 ft is accompanied by increased susceptibility to hallucinatory experiences and deficits in visual perception (Pavlicek et al., 2005). Smith (2005) performed a survey listing cognitive, psychomotor and behavioural symptoms of hypoxia with Australian Army helicopter aircrew who had operated at altitudes up to 10.000 ft. The most commonly reported symptoms were difficulty with calculations (45%), light-headedness (38%), delayed reaction time (38%), and mental confusion (36%). Non-pilot aircrew (especially loadmasters) described a significantly higher number of hypoxia symptoms than pilots and reported them more often. This might be due to their greater physical activity during flight. Analysis of narrative descriptions of the survey revealed 21 incidents where aircrew felt their performance was impaired with the potential to compromise aircraft safety (e.g. aircrew who did not respond to direct questioning, increased number of mission-outcomes impacted by reduced cognitive performance).

Stages of Hypoxia

Stage	Altitude in ft	Arterial Oxygen Saturation (%)
Indifferent	0 - 10.000	95 - 90
Compensatory	10.000 - 15.000	90 - 80
Disturbance	15.000 - 20.000	80 - 70
Critical	20.000 - 23.000	70 - 60

The table above summarizes the stages of hypoxia in relation to the altitude of occurrence, breathing air, and the arterial oxygen saturation.

1. *Indifferent Stage.* There are no observed impairments. The only adverse effect is on visual dark-adaptation, emphasizing the need for oxygen use from the ground-up during night flights.

2. *Compensatory Stage.* The physiological adjustments that occur in the respiratory and circulatory systems are adequate to provide defence against the effects of hypoxia. Factors such as environmental stress or prolonged exercise can produce certain decompensations. In general, in this stage there is an increase in pulse rate, respiratory minute volume, systolic blood pressure, and cardiac output. There is also an increase in fatigue, irritability, headache, and a decrease in judgment. The individual has difficulty with simple tests requiring mental alertness or moderate muscular coordination.

3. *Disturbance Stage.* In this stage, physiologic responses are inadequate to compensate for the oxygen deficiency, and hypoxia is evident. Subjective symptoms may include headache, fatigue, lassitude, somnolence, dizziness, "air-hunger," and euphoria. At 20.000 ft, the period of useful consciousness is 15 to 20 minutes. In some cases, there are no subjective symptoms noticeable up to the time of unconsciousness. Objective findings include:

   a. *Special Senses.* Peripheral and central vision are impaired and visual acuity is diminished. There is weakness and in coordination of the extra-ocular muscles and reduced range of visual accommodation. Touch and pain sense are lost. Hearing is one of the last senses to be affected.

   b. *Mental Processes.* The most striking symptoms of oxygen deprivation at these altitudes are classified as psychological. These are the ones which make the problem of taking corrective action so difficult. Intellectual impairment occurs early, and the pilot has difficulty recognizing an emergency situation unless he is widely experienced with hypoxia and has been very highly trained. Thinking is slow, memory is faulty, and judgment is poor.

c. *Personality Traits.* In this state of mental disturbance, there may be a release of basic personality traits and emotions. Euphoria, elation, moroseness, pugnacity, and gross overconfidence may be manifested. Behaviour may appear very similar to that noted in alcohol intoxication.

d. *Psychomotor Functions.* Muscular coordination is reduce and the performance of fine or delicate muscular movements may be impossible. As a result, there is poor handwriting, stammering, and poor coordination in flying. Hyperventilation is noted and cyanosis occurs, most noticeable in the nail beds and lips.

4. *Critical Stage.* In this stage of acute hypoxia, there is almost complete mental and physical incapacitation, leading to rapid loss of consciousness, convulsions, and finally in failure of respiration, and death.

Time of Useful Consciousness

The Time of Useful Consciousness (TUC) is that period between an individual's sudden deprivation of oxygen at a given altitude and the onset of physical or mental impairment which prohibits his taking rational action. It represents the time during which the individual can recognize his problem and re-establish an oxygen supply, initiate a descent to lower altitude, or take other corrective action. TUC is also referred to as *effective performance time.*

TUC is primarily related to altitude, but is also influenced by individual tolerances, physical activity, the way in which the hypoxia is produced, and the environmental conditions prior to the exposure. Average TUC at rest and with moderate activity at various altitudes are shown in the following table (Table - Time of Useful Consciousness). It is important to note, however, that the data present in this table are derived from a study where subjects were breathing oxygen through a mask, and that the hypoxic environment was produced by disconnecting their masks. The time of useful consciousness when hypoxic conditions are produced for an individual breathing normal air is shorter than if he had been breathing oxygen. This is because the p_AO_2 in his lungs drops immediately to a level dependent only on the final altitude, rather than dropping gradually with each breath of air, depending on lung volume, dilution of that volume, and altitude.

Time of Useful Consciousness

Altitude (1000 ft)	Rapid disconnect (moderate activity)	Rapid disconnect sitting quietly
22 ft	5 minutes	10 minutes
25 ft	2 minutes	3 minutes
28 ft	1 minute	1 minute 30 sec.
30 ft	45 seconds	1 minute 15 sec.
35 ft	30 seconds	45 seconds
40 ft	18 seconds	30 seconds
65 ft	12 seconds	12 seconds

[1]) Myocardial ischaemia or ischaemic or ischemic heart disease (IHD), is a disease characterized by reduced blood supply to the heart muscle, usually due to coronary artery disease (atherosclerosis of the coronary arteries) and is more common in men and those who have close relatives with ischaemic heart disease.

CHAPTER 4

SURVIVAL ASPECTS

The accident was not survivable for any of the aircraft occupants. The remains of the victims were removed from the wreckage area by fire fighters of the Fire Corps Special Rescue Forces. Most of the victims were found strapped in their seats. The seats had broken out of their rails in the aircraft floor during the impact sequence.

Fire

The investigation revealed no evidence of an in-flight fire. A post-crash fire developed after the aircraft impacted the ground. The parts of the vertical stabilizer, horizontal stabilizer, elevators and rudder that separated early in the impact sequence showed no evidence of fire damage.

A fire erupted between the empennage and main wreckage area, a few meters below the empennage position. The fire spread in the forest hill area in front of the empennage. A patrol fire truck was near the accident site and as a result the first emergency responders (fire fighters) arrived at the site within a few minutes of the accident. The Fire Corps trucks arrived at the accident site around 09:42 h. The firefighters applied water and aqueous film-forming foam to the fire and extinguished the fire within 10 minutes. The suppression operation was supported by Fire Corps aerial means. A Skycrane S-64E helicopter arrived first at the site and dropped water. The helicopter was based at a military airport located 3 km from the accident site. Fixed wing amphibian CL-215 and CL-415 aircraft dropped water and foam in the surrounding forest area. Firefighters then suppressed smaller fires by applying water in the forest hill area. Half an hour after the first fire was extinguished, a new fire started in the slope of the second hill. This fire spread and eventually consumed parts of the wrecked fuselage and the main wreckage area. The fire-fighting effort continued for the rest of the day. The last fire was extinguished in the late afternoon.

The fire-fighting effort continued the following day for precautionary reasons.

Emergency Response

The 199 Fire Corps National Coordination Center received a call from a firefighter in a patrol fire truck located near the accident site, who reported hearing an exploding sound near the Grammatiko hills while an aircraft flew low.

Because of the continuous information reports from the F-16s to the National Command Centers, the Fire Corps National Coordination Center had responded before the accident by alerting all the available assets in the area, including the aerial fire-fighting means from the nearest airport bases, to the highest level of readiness.

At 09:35 h, after receiving notification of the location of the accident site, the Fire Corps National Coordination Center dispatched all the available assets in the area, including aerial means and Fire Corps Special Rescue Forces. The Civil Protection National Command Center dispatched ambulances and heavy trucks (bulldozers). The General Secretariat of Civil Protection coordinated all the assets from the emergency response agencies. The Armed Forces National Command Center dispatched Air Force Rescue and Ambulance helicopters. A military ambulance helicopter landed at the accident site.

The fire trucks were unable to proceed directly to the main wreckage area because of the hilly terrain. The access road to the site was a narrow (one-lane) dirt and stone road, which had also been blocked by an overturned fire truck. As a result, the trucks had to travel in the opposite direction to another access road and then turn towards the accident site. Emergency responders had to walk to the main wreckage area through steep terrain and dense vegetation while a bulldozer cleared the road.

CHAPTER 5

TESTS AND RESEARCH

Cabin Pressure Controllers

There were two cabin pressure controllers in the aircraft. They were located on the left side of the E & E compartment, adjacent to each other on the same rack. The No. 1 unit on the left side was the master and the No. 2 unit on the right side was the slave unit. The cabin pressure controllers were manufactured by Micro-Nord in Frankfurt, Germany. The cabin pressure controllers were found at the accident site in the following condition:

- The No. 1 controller was observed to have heat damage, and its plastic BITE panel was melted. However, the pressure ports were intact.

- The No. 2 controller was slightly crushed, its plastic BITE panel was not damaged and both pressure ports were intact.

The cabin pressure controller contained a Non-Volatile Memory (NVM) chip that recorded pre-determined data and flight events during flight. The NVM contained data from 42 records, which was the maximum capacity of the NVM.

The cabin pressure controllers were transported to Nord-Micro in Germany, the cabin pressure controller manufacturer, for a NVM data download and data analysis.

The NVM Chip from the No. 2 Controller

The cabin pressure controller was examined and opened at the facilities of Nord-Micro. It was confirmed that the controller was the No. 2 unit, and installed in the slave position. The controller was opened and the NVM chip was extracted. The data on the NVM chip were copied to a new NVM chip, which was then soldered to the board of an engineering controller. The data were then extracted and analyzed using Nord-Micro hardware and software.

All the data and information related to the pin-programmed settings matched the known configuration of the accident aircraft. The oldest flight leg count was 8910 and the latest read 8984 (internal flight controller counting), which represented a total of 74 flight-legs. The elapsed time (internal counting of total time; the controller was powered with 28V DC) of the recordings on the NVM ranged from 16 384 h, 25 min, 55.4 sec to 16.686 h, 9 min, 40.7 sec, representing a time interval of 301 h, 43 min, 45 sec.

The NVM recording showed that on the accident flight, the cabin pressure control system was being operated in the manual mode. The aft OFV was constant at a 14.6 degree opening angle as measured from the fully closed position, and the flight mode was CLIMB. This was also approximately the position of the outflow valve actuator found at the accident site. Cruise flight level was selected to FL340 and the landing field elevation to 350 ft. The Cargo Heat Valve (also known as Forward Outflow Valve) and both Pack Valves and Bleeds were indicated NOT CLOSED.

The recorded data for aircraft altitude, cabin altitude, aircraft rate of climb, cabin rate of climb, and outflow valve position on flight leg 8984 (which was the last flight) were as follows:

- At an aircraft altitude of 12.500 ft, the cabin altitude was 10.000 ft. The aircraft rate of climb was 2.939 ft / min and the cabin rate of climb was 2.003 ft / min with a constant aft OFV opening angle of 14.6 degrees as counted from the fully closed position. The differential pressure (ΔP) was 0.936 psi and the flow of air through the aft OFV was 41.10 lb/min.

- At an aircraft altitude of 16.650 ft, the cabin altitude was 13.500 ft. The aircraft rate of climb was 2.831 ft / min and the cabin rate of climb was 2.304 ft / min with the valve angle 14.64 degrees constant. The differential pressure was 1.05 psi and the flow of air through the aft OFV was 40.50 lb / min.

- At an aircraft altitude of 18.200 ft, the cabin altitude was 14.700 ft. The aircraft rate of climb was 3.306 ft / min and cabin rate of climb was 2.523 ft/min. The outflow valve angle was constant at 14.54 degrees open from the fully closed position. The differential pressure was 1.117 psi and the flow of air through the aft OFV was 39.95 lb / min.

Note.- The air flow rates through the aft OFV are calculated values derived from the recorded outflow valve position and differential pressure.

There was no indication of any cabin pressure control system related malfunction or problem in any of the data recorded on the NVM.

The NVM Chip from the No. 1 Controller

The No. 1 cabin pressure controller exhibited fairly severe fire damage. The controller was examined and opened. However, data from the chip could not be read out because of its condition due to heat damage. The chip was sent to the German Federal Bureau of Aircraft Accidents Investigations (BFU) for special processing and, if positive, was to be sent back to Nord-Micro in Frankfurt, Germany for data downloading and analysis. However, these repeated attempts to obtain data from the NVM chip on the No. 1 controller were not successful.

Cabin Altitude Calculation

According to the NVM data, the cabin altitude of flight HCY 522 during the climb was as follows:

- At flight level 12.500 ft, the cabin altitude was 10.000 ft;

- At flight level 16.650 ft, the cabin altitude was 13.500 ft; and

- At flight level 18.200 ft, the cabin altitude was 14.700 ft;

According to the FDR data, the cabin altitude warning horn sounded 2 – 3 seconds before the NVM data recorded a cabin altitude of 10.000 ft. Similarly, the FDR recorded a MASTER CAUTION at an aircraft altitude of 17.000 ft (the cabin altitude was approximately 13.800 ft). The MASTER CAUTION was probably triggered by either the passenger oxygen masks deployment or illumination of the equipment cooling light.

The average aircraft rate of climb was 3.030 ft / min and the average cabin rate of climb was 2.300 ft / min. The outflow valve angle was constant at 14.54 degrees open (about 12% open) as counted from the fully closed position, and the cabin differential pressure (Δp) was 1.034 psi.

The NVM recorded a number of parameters when a specific event or condition triggered such a recording. On the accident flight, there was no other specific event or condition that would have triggered a recording, and therefore the NVM did not contain any additional aircraft and cabin altitudes. Therefore, when the aircraft had reached FL340, the cabin altitude was not recorded on the NVM.

Based on calculations by Nord-Micro, the final cabin altitude was estimated by extrapolation to have reached approximately 24.000 ft. Based on similar calculations, the Boeing Company estimated the final cabin altitude to have been between from 20.500 ft to 28.200 ft. The calculations took into account that:

- The aft OFV was open (14.54 degrees);

- The forward OFV remained open because the aft OFV remained more than 2 ±1.5 degrees from the fully closed position and, therefore, never signaled to the forward OFV to close; and

- The automatic flow control valve was assumed open because the differential pressure never reached the FCV upper limit of 1.1 psi (Δp).

Finally, any other leakages from vents, galleys, toilets, seals only served to further increase the loss of air from the cabin. The aft OFV never modulated to compensate the outflow of air because the cabin pressure controller mode selector was in the MAN (manual) position.

Previous Pressurization Leakages

The NVM chip in the cabin pressure controller recorded messages which indicated a continuous leakage situation with the aircraft that likely persisted for the last 74 flight legs. The recorded fault messages were *"30 inflow! leakage".* This fault code was recorded when the OFV was below 3 degrees opening angle for more than 5 consecutive seconds and was therefore barely able to keep the cabin pressure on schedule. These repeated messages can be explained by exceptionally low flow through the OFV due to low inflow or high leakage rates elsewhere in the fuselage.

All of these messages were recorded immediately after take off and they had a high number of reoccurrence during the same flight as shown by the intermittent count of four, i.e. there were four or more events. Both Pack Valves were indicated open, as was the FOV (Cargo Heat Valve). The messages could have been reviewed through the Built-In Test Equipment (BITE) – *Control Module Failure Messages.* However, the *"Existing Faults"* screen for the particular Fault Code *'030 INFLOW ! LEAKAGE"* would not show this information after the aircraft had landed. It would be necessary for the Ground Engineer to choose the *Fault History* menu to obtain the recorded fault messages. The Ground Engineer would normally not access the fault history, unless requested by the flight crew or unless required as a result of a write-up in the Aircraft Technical Log of an existing problem.

On flight leg 8973 a *"22CAS1 – Fail"* message was recorded. This indicated that the cabin pressure controller was not receiving the parameter CAS from DADC#1. This message was not recorded on the ground and did not represent a DCPCS problem or malfunction. Flight Leg 8632 showed the message *"2541BIT Entry"* which indicated a DCPCS calibration test on the ground.

Examinations at Boeing Equipment Analysis Laboratory

Several aircraft systems components were shipped to the Boeing Equipment Analysis Laboratory (EQA) in Seattle, Washington, United States for further examination and analysis. The examinations began on 17 October 2005. Following is a summary of the results of relevant examinations and analyses:

Automatic Pressurization Control and Indicators Panel (P5-6). The cabin pressure mode selector was confirmed to be in the MAN (manual) position by physical witness marks on the mode selector input shaft and by electrical continuity checks in the mode select circuitry. The manual outflow valve control toggle switch was found to be functional and operable.

Air Conditioning Pack Bleed Panel (P5-10). Both lamps (green light) in the MANUAL lighted (LED) annunciator were inoperable (electrically open) as a result of filament damage. Both LEDs were removed and sent for microscopic examination at the Hellenic Air Force Laboratory, Department of Aeronautics. The right filament had evidence of a "hot" fracture (electrically powered when broken). The left filament had no evidence of being powered when fractured.

Equipment Cooling Switches and OFF Indicator Light. Electrical continuity tests confirmed that both switches were in the NORMAL position, consistent with the position of the switches on the panel. Both of the Equipment Cooling Systems (Supply and Exhaust) were found to be functional in both the NORMAL and ALTERNATE switch positions. The respective lighted displays and lamp filaments for both switches were found to be intact. The right SUPPLY lamp filament showed minor stretch under magnification.

Flight Crew Oxygen Shut Off (Flow Control) Valve. A visual examination of the flight crew oxygen Shut Off Valve with a borescope revealed no observable internal obstructions.

Forward Outflow Valve. The Forward Outflow Valve was found to be functional and was observed to open and close fully when commanded electrically.

APU Bleed Duct Assembly, including the Flex Boot Flange. The pressure seal of the APU Bleed Duct Assembly, which included the Flex Boot Flange, had many torn areas; however it could not be determined if the tears had occurred prior to or as a result of the accident.

Flight Deck Oxygen Masks

The two flight deck oxygen masks recovered from the accident site were sent for DNA analysis to the Hellenic Police Headquarters Laboratory. The official report from the DNA analysis curried out by the DNA-STR (Short Tandenm Repeats) method stated:

- For biological material collected from four areas of the oxygen masks *"mixture of biological material possibly belonging to two different persons... the main genetic type appears to match that of the (First Officer)... As far as the second person's identity is concerned... we cannot express any further speculation."*

- For biological material collected from another mask area *"appears to match the type of STR of the First Officer's biological material... but, because of the fact that only two genetic areas are available, the frequency of occurrence of the particular STR in the Cypriot population is only 1/172, a fact that constitutes simple evidence that the biological material that was isolated from this mask area was from the First Officer."*

- For biological material that was isolated from another mask area *"is too restricted for us to be able to make any further statements."*

Following the DNA analysis, the masks were sent to the manufacturer, Zodiac-Intertechnique in France, for further examination and functional testing. Functional tests could not be performed due to excessive damage and presence of substances (grease, oil, soil) which might not be compatible with high pressure oxygen use.

The masks were disassembled for specific examination of the aneroid (bellows) and of the combined diaphragm. Both of these were found in good condition.

Visual inspection of No. 1 mask:

- Mask was very dirty: presence of dust, oil and soil.
- Mask was broken: right plates, harness, mask shell, main casing and cover.
- Oxygen hose and microphone cable were missing.
- Regulator was set to 100%.

Visual inspection of No. 2 mask:

- Mask was very dirty: presence of dust, oil and soil.
- Mask was broken: harness, mask shell.
- Oxygen hose and microphone cable were missing and cut at approximately 40 cm from the regulator.
- Regulator was set to 100% and emergency mode.

Accident Flight Simulation

On 30 September 2005, members of the Board and the Team, along with a representative of the Cypriot Air Accident and Incident Investigation Board (AAIIB) and a representative of the aircraft manufacturer participated in a simulation.

The simulator was owned by Olympic Airways and operated by Olympic Airlines. The latest design performance data of the simulator were the B737-300 aerodynamic data Boeing Doc-D6-58123-1 ADV. FLT.

The purpose of the simulation was to replicate the data downloaded from the NVM chip of the cabin pressure controller (aircraft altitude, cabin altitude, differential pressure, combined with various warnings concerned), and the data downloaded from the FDR.

The pressurization system of the simulator was not identical to that of the accident aircraft. The simulator's pressurization system was an analog type Cabin Pressure Control System (CPCS) and had five pressurization mode selector positions (CHECK, AUTO, STBY, MAN AC, and MAN DC). The accident aircraft had a digital type CPCS system with three mode selector positions (AUTO, ALTN, and MAN).

The pressurization mode selector on the pressurization panel was set, before the simulated flight, in the manual DC position. This position was used because the DCPCS system used a DC motor.

The toggle switch was used to place the OFV approximately 14.54° open from the fully closed position, as indicated by the OFV position indicator on the overhead panel. After takeoff no further actions were taken on the pressurization panel during the climb to FL340.

The observations/findings of the simulated flight are summarized in the following table:

Aircraft pressure Altitude	Cabin pressure altitude	Differential Pressure	Outflow Valve Angle	Aircraft Rate of Climb
12.100 ft	10.000 ft	1.0 psi	14.54°	2500 ft/min.
18.200 ft	14.600 ft	1.0 psi	14.54°	2500 ft/min
34.000 ft	26.600 ft	1.0 psi	14.54°	2500 ft/min.

All aural and visual warnings concerning cabin altitude, master caution, passenger oxygen masks, equipment cooling, were verified. The average rate of climb of the cabin altitude was approximately 2 000 ft/min.

Re-enactment of the Accident Flight

A flight was conducted on 19 December 2005. It departed Larnaca Airport three hours later in the morning than HCY 522 had departed. This was done to achieve, as close as possible, the same daylight and atmospheric conditions. Thus, instead of 06:07 h, the reenactment flight departed Larnaca at 09:07 h.

On board the B737-300 aircraft of Olympic Airlines that was used for the re-enactment were members of the Board and the Team, the Cyprus AAIIB, as well as the Hellenic District Attorney in charge of the State Judiciary Authority. The Board's medical doctor brought medical equipment for precautionary reasons given the non-standard conditions that were planned for the re-enactment flight. Portable oxygen cylinders were also available for each person on board.

During the transfer flight from Athens to Larnaca, the flight deck door was operated using normal and emergency entry procedures, in order to verify the sounds recorded by the CVR. This confirmed that the sounds recorded by the CVR at the time the aircraft was in the KEA VOR holding (a few seconds before the No.1 engine flamed out due to fuel starvation) were the sounds of the steward applying the normal and emergency procedures to enter the cockpit, and the flight deck door opening, sitting at the Captain's seat and hissing sounds similar to those of removing the mask from its stowage box and the inflation of harness and drawing of oxygen.

At Larnaca Airport, the aircraft was parked at the same apron stand as flight HCY522 on 14 August 2005 and in the same orientation, in order to replicate ambient light conditions. The pressurization panel lights, and specifically the green MANUAL annunciation, were observed and photographed using both the BRT (bright) and DIM positions of the master lights Test and Dim switch.

Before departure of the re-enactment flight from Larnaca to Athens, the air conditioning and pressurization system panels of the aircraft were preset according to the data from the NVM of the cabin pressure controller. The pressurization mode selector was set to the MAN (manual) position. The pointer of the aft OFV indicator was set in a position corresponding to approximately 15° opening position of the aft OFV, in order to maintain the cabin differential pressure of 1 - 1.2 psi during take-off and climb.

After take-off, the aircraft climbed with a rate of 2 500 ft/min and the cabin followed at almost the same rate. The aircraft leveled off at 10.000 ft, where the cabin altitude reached 8.000 ft. Up to that moment, the cabin differential pressure did not exceed approximately 1.2 psi. At 10.000 ft, and after a video recording of the instrument panels, the pressurization system was restored by turning the mode selector to AUTO, thus allowing the valves to modulate and the cabin differential pressure to be restored. Then the aircraft continued to climb to FL340.

The flight followed the same route as HCY 522 towards Athens International Airport. When the KEA VOR was reached, the aircraft entered the holding pattern at FL340 where it was intercepted by a two-seater F- 16 aircraft. The pilot sitting in the back seat of the F- 16 recorded the remainder of the re-enactment flight using a video camera.

In accordance with the observations of the F-16 pilot on 14 August 2005, two persons wearing passenger oxygen masks were placed on each side of the aircraft cabin in the over wing section. Also, the Captain of the re-enactment flight left his seat and, after a couple of minutes, a person dressed as the steward of the Operator entered the cockpit and sat in the Captain's seat.

Then the Captain returned to his seat and the aircraft left the KEA VOR holding. The aircraft followed the headings and altitudes of the accident aircraft and continued until it passed over the crash site.

The results of the re-enactment flight proved to be identical to the data downloaded from the CVR, FDR, and NVM of the accident aircraft. The F-16 pilot confirmed that his observations during the re-enactment flight were the same as those of the accident flight. The view recorded from the F-16 camera allowed the Team to verify the F-16 pilot's observations of cabin and cockpit movements inside the accident aircraft.

Mobile Telephones

There were 15 mobile telephones recovered from the accident site. Twelve were in functional status and three were found non-functional. An examination was performed of the functional sets in order to disclose:

- The owners

- Calls performed

- SMS messages

- MEDIA data (photos, videos, messages)

From the above process no data related to the accident flight were obtained. The process to obtain information was performed after the approval of the District Attorney in charge and in accordance with the existing legislation.

CHAPTER 6

ORGANIZATIONAL AND MANAGEMENT INFORMATION

Operator

General Information

The Operator was established in 1999. At the time of the accident, the Operator was operating four aircraft (one Boeing 737 - 300, two Boeing 737 - 800 and one Airbus A319 (on a wet leasing agreement)). The Operator conducted flights out of Larnaca, Cyprus and Paphos, Cyprus. During the month of August 2005, the Operator flew to 28 destinations in 11 countries.

The Operator's mission statement as stated on the carrier's website was: *"To operate a profitable low-fares airline that will provide to its passengers the highest possible safety and superior level of service."*

According to the Operator's Managing Director's message, also published on the website:

"Helios is a young airline, but it is shaped by a team of professionals with years of experience in the airline industry. Our first flight with the newly delivered Boeing 737-800 Next Generation aircraft was in May 2001 and since then we have been active

in both the charter and scheduled markets. As the first independent airline of Cyprus, our goal is to provide competition in the market and give air-travelers to and from Cyprus a choice of airline. Our aim is to provide you with safe, low-cost, convenient, and punctual air travel, packaged with a superior service – every time you fly. Our presence in the market has already driven airfares downwards. You can now fly on Helios Airways at probably half the price, after we commenced our scheduled operations. We want to persuade you to make Helios Airways our first choice airline and we are committed to providing a high quality service that is finely tuned to your traveling needs."

The owner of the airline stated that his goal was to transport passengers for Libra Holidays, a large UK tour operator. He had acquired the air carrier in 2004. He said that, at the time of the accident, the Operator was already operating approximately 68 % of its flights as scheduled services, and the remainder was non-scheduled service.

Based on records made available to the Board by the Operator, at the time of the accident the Operator had 228 corporate, administration, and line operational employees in Cyprus. The latter category consisted of 41 pilots, 104 cabin attendants, 15 ground engineers, and 15 ground operations personnel. An analysis of the Operator's employees based on their employment contract revealed that one third were seasonal employees, i.e. they did not have permanent contracts with the Operator (the numbers by category were: 7% of the pilots, 61% of the cabin attendants, 20% of the engineers, and 0% of the ground personnel).

An analysis based on nationality revealed that 72% of employees were citizens of Cyprus, 15% were dual citizens of Cyprus and another country, and 13% were foreign nationals. The Accountable Manager stated that staff salaries were identical, regardless of nationality.

The official language of communication at the company was English. English was very widely used in Cyprus since the country was part of the British Commonwealth.

Air Operator Certificate

The Operator had had an Air Operator Certificate (AOC) since 2000. The Operator's AOC No. CY-003, Issue No. 6 was issued by the Cyprus DCA on 31 May 2004 in accordance with JAR - OPS. The date of approval of the last revision was 16 February 2005. The AOC authorized the

Operator to transport passengers and cargo using the company-registered Boeing 737-300 and Boeing 737-800 aircraft on routes within Europe. The Operator had several special authorizations/approvals, including RVSM-EUR (Reduced Vertical Separation Minima – Europe).

Operator Management

According to the Operator's organizational chart in the company Operations Manual (Part A, Chapter 1, Organization and Responsibilities), the management structure consisted of an Accountable Manager and four Nominated Postholders (Flight Operations Manager, Technical Manager, Training Manager Standards, and Ground Operations Manager), as well as 14 other managerial positions. The duties and responsibilities of each manager were described in detail in Section 1.3 of the company Operations Manual.

At the time of the accident and according to Revision 8 of the Operations Manual, one staff position was vacant (Security Manager) and the position of the Training Manager Standards was covered by the Flight Operations Manager. Three positions had been staffed within a year prior to the accident (Technical Manager – April 2005; Flight Safety Officer (FSO) – joined the Operator in March 2004 and became FSO in December 2004; Quality Manager – June 2004). The Team interviewed company personnel to gain insight into the overall structure and management of the Operator.

Accountable Manager

The Accountable Manager (also Chief Executive Officer) of the Operator, a Cyprus citizen, had been in the position since June 2002. He was asked by the Team if any factors potentially degrading safety had come to his attention. He stated that *"the Flight Safety Officer follows the Flight Safety program and no factors giving rise to causes for accidents had been mentioned."*

He also stated that *"he had never been informed that the company had any problems that may be degrading safety."* He said *"Safety always was number one priority ... meetings were held regularly."* He believed that key company positions (Quality, Flight Operations, Maintenance) were appropriately staffed and managed and stated that a *"Quality System at the airline assesses productivity of the various departments with respect their safe operation."* According to the UK CAA, the Accountable Manager had been advised in November 2003, March 2004 and July 2004 of shortfalls in the standards that were expected of the Operator, and of a lack of operational management control at the airline.

When asked about the work climate at the company, the Accountable Manager stated that he had formally established an open door policy by issuing a memorandum on 20 May 2005. According to that document, he was available and willing to meet upon request with employees for 15-minute meetings not to exceed a total of two hours per week. He added that he encouraged employees to approach him, that he held meetings with an open agenda, and that the company operated *"as a family,"* including social events.

Chief Operating Officer

The Chief Operating Officer (COO) of the Operator, a British citizen, assumed the position at the company in August 2005, two weeks before the accident. In a written statement, after the accident, he said that his initial findings after joining the Operator *"gave some cause for concern."* He said *"there appeared to be a culture of fear where people were encouraged to stretch the rules to the limits"* and that *"aircraft utilization was extremely high with... inadequate downtime."* The COO said that the schedules were *"extremely tight and there was some evidence that flight times were manipulated to bring them in(to) limits."* He also mentioned that employees were prevented from taking annual leave, *"some not taking leave for 2 years."* Lastly, he said that there were no formal budgets at the company and *"decisions with regard to resourcing were not in the hands of the accountable managers."*

The COO stated that after the accident, he initiated the Operator's emergency response procedures. A consultant joined the Operator investigation team and was asked to produce a list of immediate preventive actions to be taken by the Operator. The actions he proposed included *"briefing pilots on actions in the event of decompression and the importance of air system checklist items."* The COO said that he also took action to ease the flight crew schedule and issued instructions with regard to the *"adherence to FTL (Flight Time Limitations)."* He also *"increased aircraft downtime for maintenance."*

Commercial Manager

The company Commercial Manager, a Cyprus citizen, stated that the Operator had a *"positive work climate of collaboration and mutual respect ... the current owners promote the same style of management, professional on a friendly basis."*

Nominated Postholders

According to the Operations Manual (Part A, Section 1.2.1), the Nominated Postholders must have managerial competency and appropriate technical and operational qualifications, and must be allowed to work sufficient hours (apart from flying duties) to be able to satisfactorily perform the functions associated with the operation of the company. The main functions of the management were:

- Determination of the company's flight safety policy;

- Allocation of responsibilities and duties and issuing instructions to individuals, sufficient for implementation of company policy and the maintenance of safety standards;

- Monitoring of flight safety standards;

- Recording and analysis of any deviations from company standards and ensuring corrective action;

- Evaluating the safety record of the company in order to avoid the development of undesired trends; and

- Liaison with the Authority.

The Nominated Postholders must also have:

- Expertise in the application of safety standards and safe operating practices;

- Comprehensive knowledge of international, European, and Cypriot regulations and requirements, the Operator's specifications and the need for, and the content of, the relevant parts of the manuals in their areas of responsibility;

- The appropriate management experience acquired in their previous positions of at least five years, two of which preferably acquired in an equivalent position in the aeronautical industry; and

- Familiarity with the Operator's Quality System.

Legal provisions prescribed that continuity of supervision in the absence of a nominated postholder must be ensured.

Flight Operations Manager

The Flight Operations Manager, a British citizen, stated that he had been appointed to this position in April 2003. He was asked to comment on a report that the weekly safety meetings had stopped being held after the arrival of the new Accountable Manager in June 2002. He stated that *"this is the situation I found when I was placed in this position, but meetings do take place when a situation arises every 15 days or even sooner."*

The Flight Operations Manager stated that he did not perceive any communication difficulties among crew members at the company. He was asked to characterize the accident flight crew. He stated that he had personally known the First Officer for 25 years and described him as a cheerful person with a lot of experience as a First Officer (he was one of the three senior First Officers at the company). When asked why he thought the First Officer had been unsuccessful in upgrading to Captain despite his experience, the Flight Operations Manager said that age was the main obstacle. He stated that the First Officer had moved around a lot in different companies in his career, and had thus worked for many carriers, each with its own philosophy. This had hindered him from developing important skills, particularly in the areas of CRM and decision making. The Flight Operations Manager noted that he was a good First Officer, but he just had no captaincy skills. When asked whether the First Officer had expressed any concerns about flying with the Captain of the accident flight, he denied the claims by the next of kin of the First Officer that the First Officer had repeatedly brought this issue to his attention.

The Flight Operations Manager characterized the Captain of the accident flight as a quiet person whom he had employed through a hiring agency for six months. He had had to order the Chief Pilot to conduct a line check on the Captain *"with an emphasis on CRM"* after receiving complaints. He noted that the complaints were from other First Officers (not from the First Officer of the accident flight) with low experience who had reported the Captain to be demanding. The Flight Operations Manager reported that he had asked the other two senior First Officers to comment on these complaints but they had not confirmed them. The results of the line check had been *"satisfactory."*

The Flight Operations Manager also told the Team that he had recently notified two Senior First Officers (including the First Officer of the accident flight) that the Operator was negotiating a lending arrangement with another European air carrier. If the two pilots elected to participate in the required screening test, they could be lent to the other carrier for the winter season. The Flight Operations Manager emphasized that winter demands were significantly lower than those observed in the spring and summer. The Operator was therefore looking for opportunities to keep under-occupied staff busy during the low season. The need of another airline for seasonal pilots for six months translated into good money and a good chance to gain experience on short-haul operations for First Officers whom the Operator would not have to pay to fly the reduced schedule.

Chief Pilot

The Chief Pilot, a Bulgarian citizen, had been at the position since December 2000. When asked to describe the general work climate at the company, he acknowledged the existence of cross-cultural friction in any multi-national environment, such as that of the Operator. At the same time, however, he stated that he believed that there was a family atmosphere of cooperation among pilots and described the culture at the company as "open." Some First Officers, for example, he explained, had told him that the Captain of the accident flight was not following Operator SOPs. He considered these comments premature and believed it would only been a matter of time for a newly-hired pilot, such as the Captain of the accident flight, to standardize his performance to SOPs. He characterized the Captain as strict with FCOMs and CRM, who took his duties as a commander very seriously, and who handled crews nicely. He also stated that he never saw any actual reports of complaints about the Captain.

Flight Safety Officer

The Flight Safety Officer of the company, a Cyprus citizen, had been in the position since December 2004. He stated that instead of a Safety Management System (SMS) the Operator had a Flight Safety Program which he characterized similar to an SMS. The objective of the program as described in the OM, Part A, *General/Basic Flight Operations Manual* (Section 2.3.2 *Flight Safety Programme)* was to prevent accidents. The Flight Safety Officer was responsible for promoting safety by accumulating and disseminating safety information using publications, notices, operational reminders, and recurrent safety training. The Flight Safety Officer indicated that he regularly used all of the above means to carry out his task. He stated that his office managed a Flight Data Monitoring (FDM) program, also described in the OM (Section 2.3.3 *Flight Data Monitoring)*. The Team verified that the Operator was using a FDM program. As part of the Operator's safety program, the Flight Safety Officer also stated that flight crews were invited to submit safety reports using company-supplied Air Safety Reports (irregularity reports) (Section 2.3.4 *Occurrence Reporting)*. He also stated that all elements of the Flight Safety Program were integrated into the Operator's Quality System.

Quality Manager

The Quality Manager, a Cyprus citizen, moved to the Operator from another carrier in Cyprus in June 2004. The Operator's Quality Management System was outlined in the Quality Manual (Revision 1, April 2005). The duties of the Quality Manager included the *"development, establishment, implementation and management of the Quality System... verifying through quality assurance, control and audit programmes, compliance with, and adequacy of procedures necessary for safe, cost effective operation..."* The Quality Manager was supported by two auditors, one for JAR – OPS flight operations and one for Part 145 standards. The

Quality Manager was himself an auditor for JAR-OPS Subpart M.

The areas of quality inspection included flight operations, training standards for flight and cabin crew, engineering and maintenance, including Part 145, flight safety, and the Operations Manual. Responsibility for the preparation, issue, and amendment of the manuals belonged to the respective managers (e.g. Flight Operations Manual belonged to the Flight Operations Manager).

Based on the Quality Manual, the Quality Manager should prepare a 12 month Audit Plan to ensure that each aspect of the air carrier was audited yearly, as a minimum (additional audits can be performed, if necessary). An internal self-auditing mechanism performed every three months was designed to allow management to evaluate its own Quality System and *"evaluate the overall effectiveness.., and identify and correct trends, and prevent.., future non-conformities..,"*.

When asked, the Quality Manager stated that he had not perceived any weaknesses in the maintenance or flight operations areas.

Technical Manager

The Technical Manager, a Cyprus citizen, stated that he held a Bachelor of Engineering degree in microelectronics from Bournemouth University in the UK. He was working from 1998 to 1999 as a development engineer at Marilake Instruments. From 1999 to 2002, he was with Britannia Airways as design, project and technical services engineer in aircraft avionics. From June 2003, he was working as a technical services engineer for the airline Thomas Cook. From July 2003 up to October 2004, he was working for Emirates Airlines as a project engineer. From October 2004 onwards, he had been working at the Operator. He started as a maintenance controller and after six months, he took the position of the Technical Manager in April 2005 (four and a half months before the accident).

According to the Quality Manual (Section 2-2.2.3), the Technical Manager was the only auditor for Part 145 organizations contracted to the Operator. The Technical Manager was asked whether he thought the Ground Engineers available during the month of August (month of the accident) were sufficient to cover the Operator's maintenance needs. He replied that he thought the staff was sufficient for scheduled maintenance and that, should an unexpected need arise, the Operator would request additional support from Cyprus Airways.

Maintenance Manager

The Maintenance Manager, a citizen of Slovenia, stated that he studied in Belgrade post secondary school for electrics and avionic aircraft systems for commercial aircraft. He was employed by JAT from 1988 to 1998. He then came to Cyprus and he was not employed in aviation up to April 2000. At the beginning of May 2000, he joined the Operator as a licensed aircraft engineer. Whilst with JAT, he received training covering all ATA (Air Transport Association) chapters. When he was employed by the Operator, the Cyprus DCA validated his license, which was an ICAO Type 2 license covering only electrical avionics, to cover all ATA 100 chapters. Following that, ATC Lasham issued him an authorization (No. ATCL 5000) to cover all chapters, i.e. airframe, power plant and avionics. He was responsible as a Maintenance Manager to implement the approved maintenance program.

The Maintenance Manager was asked about the accident aircraft which he described as the *"best -300"* he had encountered in his 18-year career handling this type of aircraft.

Maintenance

The Operator operated in accordance with JAR OPS 1. Therefore, the maintenance activities were carried out in accordance with JAR OPS Subpart M.

In order to obtain maintenance services and to comply with the requirements, the Operator had signed a Maintenance and Maintenance Management Services Contract with ATC Lasham in the UK. The contract defined the terms, services to be provided and the responsibilities of each party. The contract specified the technical procedures and informed the departments concerned of their responsibilities and the procedures to be followed. The aircraft to which the maintenance arrangements applied were the Operator's fleet of Boeing B737-31S (-300) and B738-86N (-800) series aircraft, fitted with CFM engines CF56-3C1 and CF56-7 series, respectively.

ATC Lasham was responsible for the Base Maintenance to be performed at Lasham Airfield in UK, in accordance with the approved maintenance programme. Under the contract, C Checks were to be performed in accordance with the approved maintenance programme and ATC Lasham's procedures in document ATC/EXP/002. Certification of Base Maintenance was to be signed off by ATC Lasham approved engineers in accordance with ATC Lasham company procedures.

The Operating Base of the company was Larnaca Airport in Cyprus, and occasional other line stations, as designated and approved by the Operator and/or Cyprus DCA. All line maintenance and minor scheduled maintenance was carried out by authorized by ATC Lasham engineers of the Operator or ATC Lasham in accordance with the Part 145 approval (UK/145/00442). All engineers should meet the requirements of ATC MOE (Maintenance Organization Exposition) Part 3, Section 3.4. Certification of line maintenance, minor scheduled maintenance, inspections and defect rectification was to be signed off by the Operator or ATC Lasham engineers as authorized by ATC Lasham.

CHAPTER 7

PROCEDURES

Operators's Procedures

The Operator's Flight Safety Manual had been prepared taking into legislation and regulations as laid down by the Joint Aviation Authority (JAA). The Manual reflected the valid company policies and procedures and was provided in order to keep flight crews and cabin crews informed of current safety regulations, emergency equipment and procedures. It outlined information that would enable the staff to carry out their duties and responsibilities with the highest degree of safety. The Flight Safety Manual was written in the English language as prescribed by JAR – OPS 1.040. The Manual comprised ten chapters as follows:

1. Introduction
2. Normal Operations
3. Supplementary Normal Operations
4. Non-normal Operations
5. Systems
6. Security
7. Aviation First Aid
8. Dangerous Goods and Weapons
9. Flight Duty Time Limitations
10. Cabin Crew Safety Notices

According to the Manual, flight crews and cabin crews were to remain familiar with the information in the Manual as it related to their respective duties and the equipment on which they were qualified. One of their obligations as a crew member was to keep the Manual up to date and current. Revisions and amendments were to be inserted immediately upon receipt. It was the primary source of information for recurrent training examinations. A copy was available in the crew briefing rooms and on board the aircraft for reference during flight duty.

For cabin crews, it was subject to periodic checks by senior cabin attendants and instructors. Cabin crews were to carry the Manual at all times when on active duty.

Flight Crew Procedures

According to the Flight Crew Operations Manual (FCOM) (page NP. 11.2, T.R. No. 2, 25 July 2005), Chap. Normal Procedures – Introduction, Par. Crew Duties:

> *Preflight and postflight crew duties are divided between the Captain and First Officer. Phase of flight duties are divided between the Pilot Flying (PF) and the Pilot Monitoring (PM).*

> *Each crewmember is responsible for moving the controls and switches in their area of responsibility. The Area of Responsibility illustrations in this section show the area of responsibility for both normal and non-normal procedures. Typical panel locations are shown.*

> *The Captain may direct actions outside of the crewmember's area of responsibility.*

Based on the Area of Responsibility illustrations, the air conditioning and pressurization sections of the Overhead Panel was the responsibility of the:

- First Officer, while the aircraft was on the ground; and

- Pilot Monitoring, while the aircraft was in the air.

First Officer [1]

According to the FCOM (page NP.21.14-16, T.R. No. 2, 25 July 2005), Chap. Normal Procedures—Amplified Procedures, Par. Preflight Procedure – First Officer, First Officer preflight duties regarding the air conditioning and pressurization system, and the flight crew oxygen system were as follows:

Air conditioning panel ....*Set*

> *AIR TEMPERATURE source selector – As needed*
> *Verify that the DUCT OVERHEAT lights are extinguished*
> *Temperature selectors – As needed*
> *Verify that the RAM DOOR FULL OPEN lights are illuminated*
> *RECIRCULATION FAN switch – AUTO*
> *Air conditioning PACK switches – One switch AUTO or HIGH, one switch OFF*
> *ISOLATION VALVE switch – AUTO*
> *Engine BLEED air switches – ON*
> *APU BLEED air switch – ON*
> *Verify that the DUAL BLEED light is illuminated*
> *Verify that the PACK TRIP OFF lights are extinguished*
> *Verify that the WING–BODY OVERHEAT lights are extinguished*
> *Verify that the BLEED TRIP OFF lights are extinguished*

Cabin pressurization panel ....*Set*

> *Verify that the AUTO FAIL light is extinguished*
> *Verify that the OFF SCHED DESCENT light is extinguished*
> *FLIGHT ALTITUDE indicator – Cruise altitude*
> *LANDING ALTITUDE indicator – Destination field elevation*
> *CABIN Rate selector – Index*
> *CABIN ALTITUDE indicator – 200 feet below destination field elevation*
> *FLT/GRD switch – GRD*
> *Pressurization mode selector – AUTO*

Oxygen ....*Test and set*

Oxygen mask – Stowed and doors closed
RESET/TEST switch – Push and hold
Verify that the yellow cross shows momentañly in the flow indicator
EMERGENCY/TEST selector – Push and hold

> *Continue to hold the RESET/TEST switch down and push the EMERGENCY/TEST selector. Verify that the yellow cross shows continuously in the flow indicator*

Release the RESET/TEST switch and the EMERGENCY/TEST selector.
Verify that the yellow does not show in the flow indicator
Normal/100% selector – 100%
Crew oxygen pressure – Check

> *Veñfy that the pressure is sufficient for dispatch*

Before starting the engines, according to the FCOM (page NP.21 .23, T.R. No.2, 25 July 2005), Chap. Normal Procedures – Amplied Procedures, Par. Before Start Procedure, the First Officer was also responsible for securing the flight deck door:

> *Flight deck door.............Closed and locked*   F/O
> *Verify that the CAB DOOR UNLOCKED/LOCK FAIL light is extinguished.*

Captain

According to the FCOM (page NP.21.19-20, T.R. No.2, 25 July 2005), Chap. Normal Procedures—Amplified Procedures, Par. Preflight Procedure, the Captain did not interfere with the setting of the air conditioning and pressurization system in the preflight phase of flight. The Captain's duties regarding the flight crew oxygen mask on his side were identical to those of the First Officer.

The Before Start checklist (QRH, page NC.1, 5 December 2003) directed the crew to verify, among other things, both the oxygen and air conditioning/pressurization systems.

(item 3 of 25) *OXYGEN & INTERPHONE* ...*CHECKED* ...

(item 12 of 25) *AIR COND & PRESS........PACK(S), BLEEDS ON, SET*

Duties after Takeoff

The first item of the After Takeoff checklist (QRH, page NC.3, Normal Checklists, 5 December 2003) directed the crew to verify the status of the air conditioning/pressurization systems.

(item 1 of 4) *AIR COND & PRESS   SET*

Page 2-35 of the company Flight Safety Manual (*Chapter 2 – Normal Operations*) described the policy regarding the use of the "Fasten Seat Belts" and "No Smoking" to signal the cabin crew of the beginning of a particular flight phase.

Non-normal Procedures

According to the QRH (page NNC.2.5, 5 December 2003), Chap. Non-Normal Checklists - Air Systems (also replicated on page NNC0.7, Chap. Non-Normal Checklists— Unannunciated), the following were to be accomplished when the cabin altitude warning horn sounds or there is rapid depressurization.

CABIN ALTITUDE WARNING OR RAPID
DEPRESSURIZATION

Condition: One or more of the following conditions:

- *The cabin altitude warning horn sounds*
-
- *There is a rapid loss of cabin press with the airplane altitude above 14.000 feet.*

OXYGEN MASKS AND REGULATORS ....*ON, 100%*
CREW COMMUNICATIONS ....*ESTABLISH*
PRESSURIZATION MODE SELECTOR ....*MAN*
OUTFLOW VALVE SWITCH ....*CLOSE*
> *If pressurization is restored, continue manual operation to maintain proper cabin altitude.*

PASSENGER SIGNS ....*ON*
> *If cabin altitude is uncontrollable:*
> *PASSENGER OXYGEN SWITCH* ....*ON*
>> *Activate passenger oxygen if cabin altitude exceeds or is expected to exceed 14 000 feet.*
> *EMERGENCY DESCENT* ....*INITIATE*
> *Accomplish the EMERGENCY DESCENT checklist if the airplane is above 14 000 feet MSL and control of the cabin pressure is not possible, or cabin pressure is lost.*

According to the Operator's Standard Operating Procedures Manual (Non-Normal Procedures-Specific, 9.1.1 AIR SYSTEM, page SOP-51, 5 August 2005):

> *NNC 2 The flight crew should don oxygen masks as a first and immediate step when the cabin altitude warning horn sounds. This action is necessary to prevent incapacitation of the flight crew due to lack of oxygen, which could result in loss of control of the airplane.*

Procedures for communicating with the cabin crew in non-normal situations were found in the Flight Safety Manual, Chapter 4 - Non-Normal Operations, 3. Emergency Calls from Flight Deck to Cabin (page FSM 4-4): In the event of an emergency arising and the Commander must brief the Senior Cabin Attendant immediately.

Cabin Crew Procedures

Guidelines regarding preflight cabin crew activities could be found in the Operator Flight Safety Manual (Chapter 2 – Normal Operations, page FSM 2-2 to 2-4), according to which there was a *"legal requirement"* for the cabin crew to conduct a pre-flight safety briefing prior to boarding the aircraft. This briefing included the exchange of flight information with both the Commander (the Captain) of the flight (i.e. expected flight time, weather information, known faults with the aircraft or ground services, special procedures that will be used) and the Senior Cabin Attendant. This latter portion of the briefing included the assignment of emergency exits and associated duties to individual cabin crew members.

Guidelines regarding cabin crew communication with the flight deck, as well as the techniques to be used to establish and maintain this interaction were outlined in the Operator's Flight Safety Manual (Chapter 2 – Normal Operations, page FSM 2-39 to 2- 40

General information regarding the cabin crew members' duties and responsibilities were outlined in Section 1.7, Duties and Responsibilities of Chapter 1 of the Operator's Operations Manual (Part A). Specifically:

... In the event of an emergency situation, they proceed to their emergency stations...; the Cabin Chief immediately contacts the flight crew for instructions.

Regarding non-normal situations, the Flight Safety Manual (Chapter 4 - Non-Normal Operations, Par. 1. General, page FSM 4-1) stated:

No two emergencies can be exactly alike. Thus aircrew must use common sense and be prepared to modify standard procedures in the interest of safety. It must therefore follow that crew members who have a clear understanding and working knowledge of standard procedures and emergency equipment will be able to handle difficult situations with confidence and success....... A thorough knowledge of all emergency equipment carried along with initiative, good judgment, alertness, and vigilance will largely influence the outcome of any emergency situation.

Procedures regarding a rapid or explosive decompression were outlined in the same manual (Chapter 4, Non-Normal Operations, Par. 9. Decompression, pages FSM 4-70 to 4-73). They were preceded by a description of the physical characteristics and the physical effects of a rapid decompression. Procedures included Immediate Actions and Subsequent Actions, excerpts of which follow:

9.1. Immediate Actions Following Rapid or Explosive Decompression.
If the cabin pressure altitude rises to or above 14000 FT, the passenger fixed oxygen masks will automatically drop..
The Flight crew will carry out an emergency descent to 10000 feet or below..
Cabin crew will:
Grab the nearest available oxygen mask and place it over nose and mouth.
Sit down on the nearest available seat and fasten seat belt..

5.Remain seated until the aircraft has leveled off.
Flight crew will:

...

2. When the aircraft has leveled to a safe altitude,_the Commander must notify the Senior Cabin Attendant.

Procedures for the cabin crew to deal with cases of pilot incapacitation (mild incapacity, sudden severe incapacity, and suspected dual pilot incapacitation) were described in the Operator's Flight Safety Manual (Chapter 4, Non-Normal Operations, pages FSM 4-75 to 4-77).

Procedures for dealing with a situation of Suspected Dual Pilot Incapacitation (page 4- 77) instructed the Senior Cabin Attendant to bang on the flight deck door, and if no reply was received to use the emergency access panel to gain access to the flight deck

---

[1] According to the introduction to the FCOM, Preflight Procedure – First Officer, *"The First Officer normally does this procedure. The Captain may do this procedure as needed."*

CHAPTER 8

AUDITS OF THE OPERATOR

As part of its investigation, the Team obtained copies of Operator audits, including follow-up actions as the result of audits. All reports concerning audits of company operations that were made available to the Team were reviewed.

For many years, the UK CAA had performed audits and inspections of the airlines in Cyprus to determine compliance with Standards defined in JAR OPS 1 and Part 145 (previously JAR 145) and the associated guidance materials. The UK CAA provided inspectors for flight operations and airworthiness audits by means of contracts with the Cyprus DCA.

2003 Audit Information

The results of an audit conducted in 2003 included a finding that the Operator lacked an in-house Quality System but had outsourced it to an external agency whose services, however, were not adequate to ensure compliance with JAR OPS requirements. At that time, the inspector had noted that an in-house Quality System should be adopted as soon as possible.

2004 Audit Information

In early 2004, the Operator was making plans to purchase the accident aircraft on 16 April 2004. In the month before the purchase, and with a proposed date for start of operations with the new aircraft of 1 May 2004, the Operator applied for variation of its AOC. Until then, the AOC was valid for the two B737-800 aircraft in its inventory. The AOC variation concerned the inclusion of the B737-300 (registration 5B-DBY). Documents generated from a series of audits performed between 23 March and 1 May 2004, were made available to the Board.

A March 2004 audit report identified deficiencies in many areas, notably the updating of the Operations Manual, training files, and compliance with and recording of scheduled and permissible duty and rest times. The system in place for monitoring currency of pilot certificates, medicals, and training was characterized as unable *"to access any historical information."* The inspector suggested that the Operator review the Operations Manual, all training files, the Quality Assurance Manual, the MELs, and all guidance material. The inspector gave the opinion that senior management personnel were not logging sufficient office time. The inspector was particularly concerned with the Operator's Quality System and noted that both the planned external (by third party) and internal (by Executive Management) audits of the Quality System had not been completed or recorded.

The inspector concluded that the Operator had not addressed *"the requirement for a robust in-house quality system"* and observed that *"It was felt that many of the above findings would have been addressed/identified and appropriate action taken if a robust quality system had been in place."*

A series of separate audits conducted in April 2004 resulted in extensive and detailed comments by the two UK CAA inspectors that conducted them. A number of items in Part A and Part D of the Operations Manual, and the Boeing 737 Differences/Familiarization course remained unaddressed. The Flight Safety Manual was found to lack an aircraft layout and Safety Equipment location diagrams. An inspection showed that the accident aircraft did not have First Aid kits, instructions on how to operate the overwing exits, or active batteries in a number of torches. Noting that the target start date for the Boeing 737-300 was 1 May 2004, the inspector commented on 7 April 2004 that, *"...as the Operator has not submitted the Part B, Part C, Part D, and Cabin Crew Manual and MELs, this date is no longer realistic."*

Before the end of April, ten days away from the target start date of operations of the accident aircraft, an inspection noted two deficiencies that still needed to be addressed prior a the Proving Flight. Both deficiencies were actioned prior to Proving flight. In view of the airline's route structure, an exemption was given for seven days for the non-carriage of an Emergency Location Transmitter (ELT). The ELT was reinstalled after four days.

Various sections of manuals (MEL, Operations Manual Part A and Part B, Cabin Safety Card, Cabin Crew Manual, Flight Safety Manual, and QRH) were found unsatisfactory and were revised and re-submitted up to two times during the three-day audit. They were eventually all noted with a recommended timescale to be rectified *"before proving flight."* A review of training records found them to again be incomplete; these were also actioned after resubmission.

The inspector observed that *"... it is evident that there is a serious problem with the management of Flight Operations. The Managing Director has intimated that he will make changes to ensure that the company is compliant with the requirements. It is recommended that the audit schedule be increased to 4 per year. In the meantime the DCA is advised to contact the Managing Director... with a view to receiving an undertaking that the management of the Flight Operations Department be reviewed and restructured."*

On 30 April 2004, a UK CAA inspector appeared to have reviewed all audit reports and determined that all deficiencies and recommendations had been satisfactorily addressed. After a flight inspection to review Flight Deck Security procedures and a Proving Flight, a JAR OPS AOC Variation (Issue) Recommendation (1 May, 2004; signed by a UK CAA inspector) was forwarded to the DCA of Cyprus, noting that: *"A full set of checks has been carried out and there are no non-conformities outstanding."* And, *"All requirements of JAR OPS 1 have been fulfilled and the operations approval is at float."*

In July 2004, two months after operations began, the same inspector, who completed the actions on 30 April 2004, conducted another audit. The inspector listed a number of serious findings, and noted that *"The lack of operational management control is now unacceptable as there is evidence of flight safety being compromised. The Accountable Manager must be asked to formally review his management structure and staff as clearly urgent action is required now."* According to the UK CAA, the Accountable Manager had been advised in November 2003, March 2004 and July 2004 about shortfalls in the standards expected of the Operator, and a lack of operational management control at the airline. Furthermore, according to the UK CAA, the Training Manager Standards had resigned in June 2004. The DCA had been alerted that the situation needed to be addressed without delay. The audit conducted in July 2004 confirmed that no action had been taken by the airline to fill the post. The DCA was again advised that the situation was unacceptable.

The report of the July 2004 audit included findings that management pilots were again found to have inadequate office duty time, the training records reviewed were incomplete, warnings of minimum number of employees' days off violations marked as such by the flight duty record program had been overridden, and the cabin crew records reviewed lacked certificates of competence for pilot incapacitation. The findings listed under the category *"Aircraft Inspection/Facilities and Organization Inspection"* were all characterized as *"...symptomatic of a lack of operational management control"* which had *"... resulted in pilots being cleared to operate public transport flights without the necessary competence, etc."* At the time of this audit, the newly-appointed Quality Manager had produced an audit program and some of the audits had already taken place.

All items in the July 2004 audit report were marked as urgent, that is, *"immediate"*, *"7 days"*, and *"before next flight"*. A handwritten note on the same report stated that *"findings in this report all finally cleared by audit 7/10 March '05"*. The note was signed 18 March 2005.

The audit report of September 2004 listed, among other things, continuing deficiencies in the areas of the Quality System, the updating of training and duty records, and the amount of office time of management pilots. The inspector also pointed out the vacancy of the Training Manager post and the fact that it must be urgently filled. This was the first report reviewed that carried comments and the signature of the head of the Safety Regulation Unit of the Cyprus DCA, who observed that *"The major issues of Training Manager Post Admin. Time for management, pilots, and quality systems improvements to be subject to follow up action."* Previous reports did not carry any comments or signatures from Cyprus DCA.

Maintenance audits were also reviewed for this time period. On 2 June 2004, ATC Lasham was audited by a UK CAA inspector, on behalf of the Cyprus DCA. The auditor noted that the findings he had identified during a previous audit in December 2003 had not been corrected. The auditor raised eight findings, including the repeat findings from the December 2003 audit. Two of the findings were considered level one and concerned control of maintenance tasks and record keeping, as follows:

- The Operator (as a JAR OPS operator) had not submitted completed maintenance documents to ATC Lasham (the maintenance contractor);

- Incomplete records of the maintenance tasks; and

- Inadequate definition of who was responsible for the completion of the Maintenance Card Summary Pages.

The Quality Manager of ATC Lasham made a proposal for corrective actions and, on 22 June 2004, the auditor clarified the two most critical findings, the Maintenance Card Summary Pages and the Airworthiness Directives Compliance Documentation. On 29 June, and 2 and 3 July 2004, further correspondence took place between the auditor, ATC Lasham and the Operator regarding the implementation of an agreed action plan. On 5 August 2004, the auditor accepted the action plan.

During a visit to ATC Lasham from 9 to 10 November 2004, by the same inspector to audit the AOC maintenance management, technical records and planning functions, the same situation was reported, including repeat level one findings. The accountability for this situation was attributed to the Operator for not complying with the obligation to forward immediately to ATC Lasham the original work package of maintenance documentation after maintenance at the Larnaca Line Station.

On 12 November 2004, the auditor, as Regional Manager, Cyprus, addressed a letter to the Technical Director of ATC Lasham stating: *"....The fact that the visit highlighted repeat findings from my December 2003 and June 2004 audits is unfortunate. Findings 1, 3, 4, 5 & 6 will be raised to level 1 if adequate corrective action has not been implemented by 17 December 2004. At this point, the Helios Airways JAR OPS Maintenance Management approval and AOC will be suspended. While it is the responsibility of Helios Airways to ensure these matters are corrected satisfactorily, ATC Lasham appears to have a contractual obligation to address findings 4, 5, 6, 7 and to monitor the action taken to correct findings 2 and 3.*

A meeting with the Helios Airways Technical Manager, ATC Lasham and the DCA is needed to review the issues raised by this audit. We would prefer these meetings to take place in Cyprus. The meeting must take place before the 3 December 2004."

In addition, the letter forwarded the following findings and comments to the Technical Director of ATC Lasham:

"Disappointing audit, several repeat findings. ATC Lasham was not able to give any real reason why the repeat audit findings had not been closed.

1. *Helios Airways does not appear to have carried out an audit on the JAR OPS maintenance management functions performed on their behalf by ATC Lasham, this year. These functions must be audited to confirm that ATC Lasham is carrying out these activities to an acceptable standard, as defined in the contract.*

2. *Numerous errors noted in completed A Check work pack documentation returned to ATC Lasham by Helios Maintenance staff at Larnaca. These include, incomplete dating of documentation, not grouping together tasks cleared by a single stamp & de-panel charts not stamped.*

3. *Helios Airways is still not returning completed maintenance documentation, including A Check Workpacks, to Technical Records, in a timely manner. These documents must be sent to Technical Records as soon as they are completed. Repeat finding noted in June 2004 Audit.*

4. *There does not appear to be a procedure for closing completed work packs of maintenance documents that are returned to Technical Records from JAR OPS operators supported by ATC Lasham. The procedure shown to me at the time of my visit was for work packs returned to Technical Records on completion of a maintenance check in one of ATC's own maintenance facilities. This did not adequately cover the circumstance described above. Repeat Finding noted in December 2003 & June 2004 Audits.*

5. It was not clear who is responsible for confirming that all

maintenance tasks have been completed on the Maintenance Card Summary Pages. On some of the records sampled, none of the tasks has been ticked off as cleared and in other cases, the summary pages had only partially been completed, although all the associated task cards were signed off. Repeat finding noted in December 2003 & June 2004 Audits.

6. Technical Records do not appear to hold the original copies of Airworthiness Directive compliance documentation ("dirty finger print copies") for Helios Airways as defined in Schedule 2 of the Helios/ ATC Contract. Repeat finding noted in December 2003 & June 2004 Audits.

7. Technical records do not appear to hold the original copies of Technical Log Pages or JAA Form 1's as defined in Schedule 2 of the Helios/ ATC Contract.

8. It is suggested that A TC Lasham introduce a work pack product sample process in order to assist in identifying some of the problems identified above.

9. Aircraft Survey of 5B-DBI was carried during visit. The aircraft was undergoing a C4 check in A TC Lasham 's hangars. Minor findings only recorded on aircraft Survey form. Form handed to Helios technical representative for action.

Annual reliability meeting held during the visit. Good meeting. For details, see Report."

The Technical Administrator of ATC Lasham responded by e-mail on 15 November 2004 to the auditor's letter expressing his concerns in relation to ATC Lasham's contractual obligations to Helios Airways concerning original maintenance documentation, as follows: "... *I have written new procedures for check clearance which should, as long as Helios operates and sends us the data, close most of your findings. I know it is no excuse but one can drag a horse to water but you cannot make it drink. ... Equally so you can request data but it does not mean they will send it and after a time one basically gives up asking. I know this is not the correct attitude or manner in which we should conduct ourselves but after a time you do give up – as I did. ...".*

The Board also reviewed a letter (unsigned) from the Director of the DCA to the Accountable Manager of Helios Airways dated 12 November 2004, regarding the 9 to 10 November 2004 audit of ATC Lasham. The Director wrote *"It is of major concern to the DCA that Helios Airways has not carried out any quality monitoring of ATC Lasham this year, as required by JAR OPS 1.900. Several serious repeat findings were noted during our audit, that according to JAA procedures should be classified as level one findings. A level one finding leaves the Department with no option but to suspend your Air Operators Certificate pending the correction of the problem, if it is to remain in compliance with JAA requirements. However, the findings have been temporarily classified as level two findings in order to give Helios some time to correct them. It must be confirmed by 17 December 2004 that all of our audit findings have been corrected. If this has not been achieved, the open findings will then have to be reclassified as level one."*

According to the UK CAA, the UK CAA inspector visited ATC Lasham on 7 December 2004 and met with the Technical Managers of ATC Lasham and the Operator. At this meeting, all outstanding findings were confirmed closed and a letter dated 9 December 2004 was sent to the DCA, Cyprus confirming this

2005 Audit Information

There was evidence of two audits having been conducted in 2005, prior to the accident. At that time, the Operator had a new owner. The Operator temporarily staffed the position of Training Manager Standards with the Flight Operations Manager, who was now managing double duties. According to the Accountable Manager's statement, a new employee had been selected for that post but was injured in an accident. The Operator had to cancel plans to hire him.

The Board was not supplied with the report for the first audit conducted in March 2005. Instead, the findings were listed in a letter written by a UK CAA inspector to the Operator's Flight Operations Manager noting the results from the *"Routine Regulatory Inspection"* he conducted between 7 and 10 March 2005. The letter specified that the Quality System and the

Quality and Operations Manuals were found to continue to have serious deficiencies and the recommended timescale to address specific issues was defined as two and three months. The inspector reported that *"Not all company personnel had been provided with quality related briefing as required....".* He also reported that *"The flight and ground operations managers and engineering and maintenance manager had not attended the* (quarterly management quality review) *meetings as required."*

The inspector wrote that issues related to the training records reviewed were expected to be addressed by the new to-be-hired Training Manager Standards (date not noted). Other areas of concern were the capture and processing of Commander Discretion Reports (variation from flight time limitations), and of training. A review of the Operator's organization and management led the inspector to note that *"... the organization and management of Helios Airways and associated operational supervision was not properly matched to the scale and scope of the present operation."*

The UK CAA inspector concluded that *"... the main concern... is that of a lack of resources, which result in the not having the relevant and related staff in certain areas of operation."* The inspector ended by advising that a follow-up visit take place from 27 to 28 April 2005 and he signed the letter, adding, *"For Director of Civil Aviation."*

There were no records provided for any audits having taken place in April 2005. The next audit report reviewed was from June 2005. In that report, the same inspector, who conducted the March 2005 audit, noted that *"...there appears to be an improvement in management's attempts to address issues raised by regulatory audit. And too there appears to be some small improvement in resources, but it remains to be seen as to how long such improvements continue, and whether or not further improvements will follow."*

Issues with the capturing and processing of Commander Discretion Reports, training records, and aspects of the Quality System were noted, similar to those noted in the March 2005 report. The June 2005 report was also signed by the Cypriot Flight Operations Officer of the Safety Regulation Unit of the Cyprus DCA, who recommended continuance of the Operator's AOC.

Two reports from an audit conducted in September 2005, about two months after the accident noted, among others, four issues. The first three issues were identical to those noted in the previous audits: Insufficient office time of management pilots, lack in robustness of the

Quality System, and an impractical method of scheduling timing of routes to ensure adherence to Flight Time Limitations. The fourth issue specified that *"The normal checklist did not appear in any manual aboard the aircraft."* The aircraft referenced was the B737-800.

The Board also reviewed maintenance audit findings for 2005. A JAR OPS Subpart M audit by the Cyprus DCA was carried out from 8 to 9 June 2005. All non-conformances were considered Level 2. Helios Airways was instructed to inform the DCA by 31 July 2005 of the actions taken to address the findings. The Helios Maintenance Technical Director responded on 2 August 2005.

A Line Station audit in Larnaca was carried out in June 2005 by the ATC Lasham Quality Manager. Nine findings were recorded concerning several technical operations matters, such as manpower planning, processing matters, documentation, and material and equipment management.

ATC Lasham

ATC Lasham Limited was the maintenance contractor for Helios Airways, situated at Lasham Airfield in the UK. It was a Part 145 maintenance organization approved by the United Kingdom Civil Aviation Authority (under EC Regulation 2042/2003 Annex II Part 145), to maintain the products listed in the approval schedule and issue related certificates of release to service using the reference: UK. 145.00442.

Lasham Airfield had been an operational maintenance base since 1953 when the site was occupied by Dan Air Engineering Ltd, and more recently by FLS Limited. Over the years, the personnel at the base had remained the same although company names had changed. In 1994, Aviation Tool Corporation acquired the facility and formed ATC Lasham using the same core personnel.

In October 2002, ATC Lasham acquired the maintenance facilities of Heavylift Aircraft Engineering Ltd at Southend Airport. The company had performed base maintenance on a wide variety of aircraft over the years. There were four hangars capable of accommodating Boeing 757 and Airbus aircraft. ATC Lasham held the approval of the Civil Aviation Authorities of 16 countries, among them the Republic of Cyprus under reference: DCA 002 /2000.

Department of Civil Aviation in the Republic of Cyprus

A State's responsibilities under the *Convention on International Civil Aviation* (Chicago, 7 December 1944) include the licensing of operational personnel, the certification of aircraft, air operators and maintenance organizations. In order to discharge its responsibilities, each State should enact a basic aviation law which will provide for the development and promulgation of a code of air navigation rules and regulations which should be consistent with the provisions of the Annexes to the Convention. The regulations should provide a framework of positive control and guidance. The State regulations should require the operator to submit detailed information on the organization, method of control and supervision of flight operations, training program and maintenance arrangements as a basis for operational certification. The operator's material would normally be submitted in the form of an operations manual, a maintenance control manual and a maintenance program.

Continuing surveillance by the State of an air operator certificate holder's operations is inherent in the system of certification and it is an essential part of the State's responsibility to ensure that the required standards of operations are maintained. Adequate authority for certification and continuing surveillance of an air operator certificate holder's operations should be contained in the provisions of the basic aviation law of the State, and adequate resources must be allocated for this activity.

Safety oversight of air operations in Cyprus rested with the Department of Civil Aviation (DCA). The Board examined the organization, and means by which the Cyprus DCA was able to and performed its duties as outlined by the *Convention on International Civil Aviation* and in the Annexes thereto.

The Safety Regulation Unit (SRU) was part of the Cyprus DCA, which operated under the Ministry of Communications and Works. The DCA conducted safety oversight duties based on the Cyprus Aviation Law (Law 213(I)/2002). The most recent amendment (2004) to the Law included the transfer of regulations from the JAA to EASA.

The SRU consisted of three Sections (Operations, Airworthiness, and Licensing) which together had safety oversight responsibility over three local air carriers with DCA-issued AOCs (17 aircraft in total), as well as 60 General Aviation aircraft, 212 non-Cyprus DCA licensed ATPLs, two FTOs for General Aviation training, and three AMOs (2005 post-accident data).

At the time of the investigation, each of the three Sections was staffed by no more than three technical employees (not all of which held full-time appointments). Each Section was structured to have its own Head, however, the position of one Head of Section (Operations) was vacant and the Head of another Section (Airworthiness) held a part time contract. The Unit also employed four inspectors to carry out the safety oversight duties. Two of the inspectors were Cypriot and resided in Cyprus (one was employed full-time and the other part-time). The other two inspectors were British and were selected and assigned to their posts by the UK CAA based on contractual agreements with the Cyprus DCA.

The Board collected statements from the newly-appointed Director of the Cyprus DCA and the SRU Head and employees. The DCA Director stated that the DCA was under-staffed but that a proposal at the Ministry level aimed to increase the number of employees. The SRU Head stated that his staff was supported by the UK CAA and that all oversight activities were carried out per regulations. The two staff members of the Operations Section that were interviewed confirmed that checks on Cyprus airlines were performed as required, and added that the last check on Helios Airways had been conducted in July 2005 and had shown the operations to be in compliance with JAR OPS 1 regulations. When asked for his views regarding the potential for preferential treatment towards an airline, the Cypriot Operations Inspector stated that no airline was given preferential treatment. The Airworthiness Section Head stated that all airworthiness inspections on the aircraft of Helios Airways were performed per regulation.

The Head of the Licensing Section gave a different view of the SRU. He stated that inspections were not performed as required (i.e. they were only carried out "on paper") and that he thought the Cyprus DCA displayed preferential treatment towards Helios Airways (e.g. no SAFA audit had been performed on the aircraft of the Operator for the past 10 months and it was now operating a foreign aircraft without SAFA checks). He had attempted to communicate his concerns in writing to the Minister and did not observe any changes until 11 months later when the DCA performed sample checks on all air operators in Cyprus.

The Role of the UK CAA

Formal agreements for the provision of services of UK CAA Airworthiness Surveyors, Flight Operations, and Flight Test personnel to assist the DCA of Cyprus in its oversight responsibilities over the three air carriers with Cyprus AOCs existed in the form of three separate contracts that were renewed annually. A fourth contract referred to general technical assistance and advice for operators with aircraft below 5 700 kg.

Contract Number ITCS/0003/CYPR described the services of the UK CAA in assisting the Cyprus DCA in its oversight of Cyprus air carrier operations. The scope of services in the 2003 contract (Appendix 35 to the ITCS contract) referred to the two Boeing 737- 800 operated by Helios Airways at that time and included the surveillance of the operation and maintenance of aircraft, provision of Flight Operations inspection and surveillance of maintenance arrangements, monitoring of documentation relating to maintenance and provision of continuing airworthiness information, and flight testing for renewal of Certificate of Airworthiness. The scope of services in the 2004 contract (Appendix 39 to the ITCS contract) was extended to include the B737-300 aircraft (registration 5B-DBY).

A list of planned and accomplished audits of the three Cyprus AOC holders for the year 2005 appeared in a letter dated 16 August 2005. According to this letter, signed by one of the two UK inspectors, and addressed to the Minister of Communications and Works of Cyprus:

"The UK CAA has, for many years, provided an advisory service to the Department of Civil Aviation (DCA) and has performed audits and inspections of the airlines in Cyprus against the standards defined in JAR OPS 1 and the associated guidance material. The operators are informed by the DCA in writing of the findings from our audits. The responsibility rests with the DCA to decide the implementation of any necessary corrective actions. The UK CAA has no legal authority to take such decisions."

Based on the schedule attached to the letter, each of the three air carriers had been or was going to be audited twice in 2005 (including both "in-depth" and "follow-up" Operations audits). Four audits of Helios Airways operations appeared to have been carried out between January and August of 2005: an in-depth operations audit (7 to 11 March), a follow-up operations audit (1 to 2 June), a JAR OPS 1, Subpart M maintenance audit (8 to 9 June), and an aircraft check (all operators on 5 August

2005). Comparing this list with the audit reports made available, the Board was able to confirm that these audits were performed. The letter concluded that the three air carriers "...*conform to the requirements of JAR OPS.*"

Audits of Cyprus DCA

The Board acquired copies of reports that referred to evaluations of the Cyprus DCA.

a. ICAO – 1996: The first ICAO audit was part of the voluntary ICAO Safety Oversight Audit Programme;

b. ICAO – 1999, 2002: The audits were conducted by ICAO as part of its Universal Safety Oversight Audit Program (USOAP); a scheduled audit in 1999 and a follow-up audit in 2002. A letter from the President of the Council of ICAO in 2005 summarized the findings from these audits;

c. JAA – 2003, 2004, and 2005: Two reports referred to the Fact Finding initial visit (22 to 26 September 2003) and the follow-up visit (25 to 26 March 2004 – draft report only) by the Joint Aviation Authorities (JAA) with the purpose of evaluating the DCA for full JAA membership;

d. EC – 2005: Following the accident and at the request of the government of Cyprus, the European Commission (EC) conducted an evaluation in August 2005; and

e. Private firm – 2005: In November 2005, at the request of the government of Cyprus, a private consulting firm conducted an evaluation of the structure (Diagnostic) of the DCA to support safety and safety oversight.

Audits by ICAO

In 1996, as part of the voluntary Universal Safety Oversight Audit Program

(USOAP), ICAO was requested to conduct an assessment of the safety oversight capability of the Cyprus DCA to confirm its conformity with ICAO Standards and Recommended Practices (SARPs). The audit was conducted by an ICAO team from 6 to 8 May 1996.

The report of the audit identified numerous deficiencies that needed to be addressed by the Cyprus DCA to bring its regulatory and safety oversight program into compliance ICAO Standards and recommended Practices. The Findings and Recommendations of the audit covered a wide spectrum of issues, including primary legislation and operating regulations, organizational structure, policies, procedures, personnel staffing and qualifications, and recordkeeping. Some of the Findings were as follows:

- Regarding Personnel Licensing: *"The DCA does not have a structure, policy, or procedures for personnel licensing and training."*

- Regarding Flight Operations: *"The DCA Director has no authority to... establish a flight operations inspection organization to assist in carrying out the functions and responsibilities of the DCA." "The DCA has limited information available regarding the system of air operator certification and surveillance." "There is no established system for the continued surveillance of operators...."*

- Regarding Airworthiness, *"The DCA does not fully comply with Article 12 of the Convention, which requires the State to assume responsibility for ascertaining that aircraft on its registry and operations within its jurisdiction comply with the Standards laid down in the Annexes to the Convention."*

Although the Cyprus DCA had arranged contracts with the UK CAA and Bureau Veritas, thereby delegating many of the safety oversight tasks to outside agencies, the Cyprus DCA was still responsible for monitoring and directing the delegated functions. The report of the audit revealed that this was not done in many areas of responsibility.

The 1999 ICAO audit was a follow-up to the initial, voluntary audit in 1996. The report concluded that the *"technical review, inspection, and evaluation tasks were conducted by the UK CAA, and the DCA was responsible for taking action on the basis of information and advice provided by the UK CAA."* but continued that *"The DCA had no internal expertise to assess the inspection conducted on its behalf or to assess any other technical aspects of the work performed."*

Specifically, the report cited the difficulties of the DCA to bring the civil aviation system into compliance with ICAO Standards and Recommended Practices (SARPs), and the insufficient and inadequately trained and qualified staffing of the Cyprus DCA. The corrective action plan included the intention to use the services of the UK CAA to develop appropriate regulations for airworthiness and flight operations matters and the services of a legal advisor to develop primary legislation and civil aviation regulations (e.g., Civil Aviation Law) (stated completion dates by the end of 1999 and 2000, respectively), and to recruit additional personnel, including inspectors, who would be properly trained to gradually function independently by the end of 2001.

Another corrective action mentioned the intention to assess the feasibility of establishing a licensing system for commercial pilots and maintenance engineers which, at that time, did not exist.

Two years later, in 2002, the follow-up audit determined *"limited progress"* in key areas. The draft Civil Aviation Law was found not to contain *"several essential provisions considered to be important for ... an effective and efficient system of safety oversight in Cyprus."* Its implementation and the formal adoption of JARs were moved forward to March 2003.

Deficiencies were again identified in the organization of the DCA, in its staffing, in the definition of duties and responsibilities, in the recruitment of qualified personnel, and in a formal training policy and programme for inspectors. Such problems were attributed in part to the *"inevitable bureaucracy"* resulting from the fact that the DCA was operating as a functional department of the Ministry of Communications and Works which *"limit its (the DCA 's) effectiveness and efficiency."* The anticipated completion of the planned recruitment and training of inspectors had not been accomplished and was also moved forward to 2003. A system for pilot and engineer licensing (issuance and validation) was still not in place.

Following the accident, in August 2005, in response to a letter from the Cyprus government, the President of the Council of ICAO addressed the Ministry of Communications and Works of Cyprus and referred to the ICAO audits. In his letter, the President expressed *"... the magnitude of our (ICAO's) concern..."* about the safety oversight capability of the DCA of

Cyprus in quantitative terms. Citing that *"A lack of effective implementation in excess of 15 per cent generally indicates significant problems in terms of State oversight capability"*, the letter indicated that *"the 2002 audit ... determined the lack of effective implementation of corrective action plans following the previous audits was 46.57 per cent."*

Audits by JAA

A November 2003 report summarized the findings of the first JAA fact-finding evaluation visit from 22 to 26 September 2003 and recommended that the DCA *"... urgently develop its ownership of the commitments made towards the Chicago Convention and the Cyprus Arrangements by ... increasing its human resources in all JARs related fields (Airworthiness/Maintenance, Operations, and Licensing) and providing the necessary training to its personnel."*

The JAA also conducted site visits to the three air operators and reported that they *"... revealed exemplary standards in every respect."* The JAA concluded that *"Continued oversight of these operators, contracted out to the UK CAA, was excellent."* The JAA appeared overall satisfied with the progress that had been made since the last visit, specifically noting that *"... the presence of the current consultant is helping to speed up the process"* but at the same time recommending that the DCA take *"... more ownership of the Aviation Safety Regulatory activity by developing its own aviation safety culture..."* and emphasizing the need for the DCA to improve its organization and pay particular attention to finding *"... a comprehensive solution to the complex personnel matter."*

The report concluded by estimating that the Cyprus DCA would qualify for full JAA membership within the next nine months (during which time the identified issues would have been addressed).

Six months later, the JAA followed up with another site visit to assess progress in addressing the issues identified after the initial visit. In its report (draft report: April 2004), the JAA noted significant progress in many areas. It concluded that the Cyprus DCA fully qualified for JAA full membership and made this recommendation official to the JAA Committee, but noted in various places that *"... the DCA must maintain its existing programme of support by external agency expert assistance..."* arguing that *"It remains ... clear that the presence of the current consultant is an essential part of the good running of the system, at least for the time being."*

In a letter to the Cyprus DCA in April 2005, the Operations Director of the JAA summarized the main findings of a January 2005 OPST visit to Cyprus and listed seven areas of SRU functions that require attention, among them:

- *"the Authority has not established a procedure for renewal of AOCs, as AOCs have no expiry date.*

- *Only one (of four) of the inspecting staff is permanently employed as a member of the Safety Regulation Unit (SRU). Two inspectors are employed by short term contracts, the remaining inspector is seconded from the UK CAA. This situation seems to be inadequate in order to fulfill the responsibilities and duties of an oversight function...*

- *... the Authority could not document how findings from inspections carried out in 2004 had been responded to by operators. The Authority could not provide an inspection plan for the previous year (2004).*

- *The Authority has not established a procedure for issuing Operational Directives."*

The letter congratulated the Cyprus DCA *"on the overall standards and level of achievement in regard to JAR-OPS and JIPS... achieved thus far"* but noted that *"Once we have received and reviewed your proposals for corrective actions, and provided they are acceptable to CJAA, we may then be in a position to recommend 'Mutual Recognition' status....".*

Evaluation by the European Commission

The 2005 EC evaluation was conducted by two officials from 23 to 26 August 2005, following a request by the Cyprus government after the accident. The Commission did not publish an official report but made general comments in a letter addressed to the Permanent Representative of Cyprus to the European Union and signed by the Director-General of the EU.

Regarding the general organization of air safety in Cyprus, the letter stated *"...in particular, the oversight function of governmental authorities with a view to preparing the more comprehensive standardization visit to be carried out by the EASA and to identify priorities for further action."* The officials found the Cyprus DCA to still be *"lacking the necessary strength to comply fully with its international obligations."*

In his letter, the Director General of the EC recommended that the Cypriot *Government "...take the necessary political commitment to supply this Department [DCA] with the resources required to carry out fully its safety oversight function and to reorganize the chain of command in order to give safety the high priority it deserves inside the organization."*

Assessment by a Private Firm

After the accident, a private consulting firm was called in for an assessment. It used the International Aviation Safety Assessment (IASA) auditor checklists developed by the United States Federal Aviation Administration (FAA) as a basis for a *"diagnostic"* assessment of the DCA.

The firm provided the DCA with a comprehensive assessment based on three criteria: the ICAO requirements contained in Annexes 1, 6, 8, 14, and 17; the JAA/EASA requirements (Safety Assessment of Foreign Aircraft); and the FAA - IASA requirements. The objectives of the examination included an evaluation of the structure (organization) of the DCA to support safety and safety oversight, and concentrated on the *"SRU's ability and capability to ensure the civil aviation system operates safely and sustains compliance with the standards and recommended practices (SARPS) of ICAO Annexes 1, 6, and 8..."* The examination included a review of non-conformities contained in technical reports related to the Cyprus DCA prepared previously by ICAO, the UK CAA, and by JAA/EASA. It was determined that:

> *"The structure of the DCA to support safety oversight is inadequate to support current and future operations;"*

> *"The Systems supporting the technical programs are not fully implemented in the areas of safety and security;"*

> *"At the time of this Diagnostic, there was no evidence that confirmed the existence of any Risk Management process within the DCA."*

Insofar the overall SRU was concerned, the report pointed out the absence of an official Administration Policy Manual outlining duties and responsibilities, job descriptions, and qualifications for individual post-holders. Staffing, the hiring process, as well as initial and recurrent training were all characterized as inadequate.

The contractual agreements between the Cyprus DCA and the UK CAA were reviewed and the report concluded that *"... the DCA is not receiving the full scope of contracted services from the UK CAA... the DCA has not taken a more controlling position in directing the contracted service provider (UK CAA) and allowed them to perform as they saw fit."*

The Licensing Section was found to be involved in activities only at the Private Pilot Licensing (PPL) Level and *"...did not include any type of verification and recordkeeping of any licenses issued by EU Member states or validation and recordkeeping of licenses issued by any other non-EU state."* Furthermore, there appeared to be *"...minimal supervision of Aviation Medical Examiners (AMEs)."*

The Flight Operations Section was found to rely on one *"...full time UK CAA Flight Operations Inspector, who only conducts ground inspections... and another UK CAA FOI that travels to Cyprus once a year to perform en-route inspections (cockpit and cabin)."* The actual inspection program was found not to follow the published plans (annual, instead of bi-annual, en-route inspections in 2004 and 2005), especially in view of the fact that the latter inspector traveled to Cyprus only once a year for two to three weeks.

The Airworthiness Section was reportedly *"...understaffed and also unable to adequately cover required ICAO safety oversight requirements...".*

Following recommendations by the private firm, the DCA established a procedure to hire qualified Engineers and Captains to the SRU to perform the civil aviation inspector duties after the required on-the-job training. Also, a "Go Forward Plan" was signed with the private firm in order to enhance and improve the safety oversight capabilities of the DCA. The compliance with the ICAO and EASA/JAA requirements was to be accomplished in the shortest possible time.

CHAPTER 9

PRESSURIZATION INCIDENTS

With the Accident Aircraft

On 16 December 2004, Helios Airways flight HCY 535 was en route from Warsaw, Poland, to Larnaca, Cyprus, cruising at FL350 at position EVENO, just prior to the top of descent to Larnaca, when the aircraft experienced a rapid decompression. The cabin altitude warning horn sounded and the cabin altitude went up rapidly. The passenger oxygen masks deployed automatically.

The flight crew initiated an emergency descent to FL100 and declared an emergency (MAYDAY) call to ATC. At 19:32 h, the Captain reported that the *"cabin state has stabilized."* The chief cabin attendant reported to the Captain that there had been a bang from the aft service door and there was a fairly large hole in the seal of the door, through which *"a hand would fit."* The flight continued to Larnaca where the flight was given assistance by Larnaca ATC for landing. Emergency services were standing by for the landing. Three passengers were admitted to Larnaca General Hospital with ear problems. The flight crew reported that the automated cabin announcement (related to the decompression and dropping of oxygen masks) was in German. The Captain stated that there might have been a problem with the outflow valve. The First Officer reported that his face mask was filled with mist during the initiation of the emergency descent. He also reported that the aircraft remained in the airway during the emergency descent (the Board noted that the aircraft should have deviated from the airway per procedure). This was confirmed by the Captain. On 19 December 2004, the FDR and CVR were removed for downloading (Technical Log book p. 652).

The Air Accident and Incident Investigation Board (AAIIB) of Cyprus was not able to reach a conclusive decision as to the causes of the incident, but indicated two possibilities:

1. *"An electrical malfunction caused the opening of the outflow valve; and*

2. *An inadvertent opening of the aft service door with the handle not in the closed position."*

In addition, the report stated that *"the Committee unveiled a number of inconsistencies and omissions"* but the report did not list these inconsistencies and omissions.

The AAIIB - Cyprus report contained three safety recommendations:

a. *"Aircraft experiencing malfunctions necessitating emergency descent should immediately clear controlled airspace by initiating a turn. ATC should ensure that other aircraft known to be in the vicinity are informed of the emergency descent as soon as practicable;*

b. *The CVR circuit breaker should be pulled in all cases following a serious incident or accident after landing so that the last 30 minutes of communication are preserved. This should be included in the pilot's SOPs; and*

c. *The Technical Log engineering responses or action taken should be clear, precise and closely related to the pilot entries in the Technical Log."*

Equipment cooling problems existed on the accident aircraft between 9 June and 14 August 2005. In this period seven incidents involving the cooling system were reported.

Pressurization Incidents with other Aircraft

Shortly after the accident involving HCY522, the Irish Air Accident Investigation Unit (AAIU) forwarded to the Team Final Reports on a number of previous occurrences involving pressurization problems.

Boeing 737-548, Irish Registration EI-CDB [1]

On 7 December 2000, the master caution light illuminated as the crew were completing the AFTER TAKEOFF checklist. The pilot cancelled the warning and advised the copilot that the "Auto Fail" warning light on the overhead pressurization panel was illuminated. The crew continued with the AFTER TAKEOFF checklist. About this same time, the senior cabin attendant entered the cockpit advising that she and another cabin attendant were experiencing problems with their ears. The crew accomplished the AUTO-FAIL non-normal checklist which directed them to select the STANDBY pressurization mode as the flight continued to climb. At FL100, the Cabin Altitude Warning horn sounded, signaling that the cabin pressure had reached 10.000 ft. The crew silenced the warning and, having already reached FL110, they requested a descent to FL100, where they continued the checklist. While performing the checklist, the pilots reported being confused about the location of the cabin altitude warning horn cutout button and location of the Cabin/Flight Altitude Placard. Neither of them checked the cabin altitude indicator. After the checklist was completed and assuming that the "Standby Mode" was operating correctly, the crew requested and was granted approval to climb to FL290.

Passing through FL125 the copilot (PF) informed the pilot that he was feeling light-headed and that he was going on oxygen. He also reported that the cabin was rising with the aircraft altitude and would soon reach FL140 where the passenger masks would drop. He did not fully don his mask but held it against his face to draw oxygen as he needed it. About this time, the pilot took over PF duties.

The senior cabin attendant again entered the cockpit to report that passengers were experiencing ear problems, the cabin was very cold, and there was some misting in the aft cabin. Shortly after, she also reported that the passenger oxygen masks had deployed. At this point, at about FL141, the pilot leveled the aircraft and made a passenger announcement. The pilot then requested descent for landing at the departure airport.

During interviews, the pilot reported that he could not fully recall the events that occurred during the initial descent from FL140. He also stated that he had never felt the need for and had thus never gone on oxygen. His first recollection was that of approaching FL070. The copilot reported that the pilot had called for descent checks on two occasions and had briefed for a CAT II approach.

The AAIU investigation revealed that the air conditioning packs were not switched on during the After Start checks and remained off until the aircraft descended towards its destination. In this condition, the cabin was never pressurized and its altitude climbed at the same altitude as the aircraft.

In its Final Report, the AAIU recognized that a number of technical and operational issues had distracted the flight crew's attention during the preflight checks and that this was not uncommon in flight operations. It was noted, however, that *The continued distractions... should have acted as a reminder to both flight crew members that there was a strong likelihood that a check may have been missed.* The AAIU concluded that *"... it is clear that both flight crew members did not adhere to the Operator's SOP's with regard to completion and verification of the normal checklist, as the packs were not selected on and the flight crew took-off under the assumption that the pressurization system was properly configured."*

The AAIU also observed that, later, at the onset of the "Auto Fail" light and the Cabin Altitude warning horn, *"...neither flight crew member carried out a thorough analysis of the situation, in order to determine the true selection and condition of the pressurization system."*

Furthermore, although *"The Captain... did not fully appreciate the significance of all the information provided to him during the unfolding of events and thus allowed the un-pressurized aircraft to continue its climb"*, the AAIU observed that *"The continued persistence of the (Senior cabin attendant) in keeping the flight crew informed of the cabin situation was a major factor in ensuring the safe outcome of this serious incident."*

Boeing 737-204, Irish Registration EI-CJE [2]

On 28 September 2002, the pilot made an engine bleeds off takeoff to demonstrate the appropriate procedure to a relatively inexperienced copilot. During the after takeoff checklist, the copilot turned the bleeds on, announcing his actions orally to the pilot as per SOP; however, he also switched the air conditioning packs off without announcing orally his actions. About FL240, the flight crew heard what they understood to be the configuration warning horn. Checks on the aircraft configuration and reference to the QRH failed to detect the reason for the warning.

The flight crew requested and received clearance from ATC to level off at FL270 to continue trouble-shooting the warning horn. When a cabin attendant called the cockpit, the pilot was busy with the problem and he advised that he would call her shortly. The pilot, initially still thinking the horn signified a configuration problem, tried to correct it by moving the throttles. When this action failed to silence the horn, he noticed that the air conditioning packs were off and he switched them back on. Because the cabin altitude had reached 14.000 feet he instructed oxygen masks on while the copilot ran the decompression checklist. In the meantime, the passenger masks had also dropped in the cabin though later the cabin crew stated they did not personally notice any signs or symptoms associated with an unpressurized cabin. Shortly, cabin pressurization returned to normal and the cabin altitude horn silenced. The crew requested and received clearance to FL310 and continued on to the scheduled destination without informing ATC of the incident at any stage.

In its analysis of this occurrence, the Irish AAIU stated that *"...the misdiagnosis of the warning horn and the time it took to solve the problem, inability to recall the maximum cabin altitude attained, the inability of both pilots to recall any warning lights, can only lead the investigation to conclude that both pilots were experiencing the onset of hypoxia."* The AAIU adds, *"...the decision to climb to FL310 without informing ATC of the onboard problem was perhaps, in part, the result of the crew's same hypoxic experience."*

The AAIU also analyzed the *"locked door"* policy regarding accessibility to the cockpit by cabin crew in the event of incapacitation of both pilots; concluding, *"While the locked door cockpit door policy, complying with ICAO standards and the Operations instructions, had no direct bearing on the outcome of the incident, the potential for a full scale accident is self evident in this type of emergency."*

Boeing 737-204, Irish Registration EI-CJC [3]

On 8 November 2004, the aircraft made an engine bleeds off takeoff and the aircraft climbed to FL320 en route to its destination. About 10 to 15 minutes after reaching cruise flight, the cabin altitude warning horn sounded. The flight crew immediately donned their oxygen masks, established communications, and completed the ALTITUDE WARNING checklist. They noticed that the cabin altitude was about 11.000 – 12.000 ft and climbing slightly less than 2000 ft/min. The EMERGENCY DESCENT checklist was called for and an emergency descent was executed.

A cabin attendant occupying a passenger seat for company positioning informed the no. 1 cabin attendant that *"...some of the passengers showed signs of hypoxia – some appeared to be dizzy and laughing and some did not bother to put on their oxygen masks."* An uneventful landing was made and no serious injuries were reported.

In its Final Report, the AAIU discussed the fact that hypoxia effects can be so subtle and insidious that *"A flight crew operating under a high workload on the flight deck may not fully appreciate or recognize the initial symptoms of hypoxia. It is therefore possible that judgment may be impaired to such an extent that corrective actions associated with dealing with an emergency situation may lead to incorrect or inappropriate response which could endanger the aircraft."* The AAIU observed that common findings in the number of pressurization and air conditioning events it had investigated over the years were *"... failure to adhere to Standard Operating Procedures (SOP's), non-use or non-accomplishment of checklists, and inadequate vigilance specific to the pressurization and air conditioning panel."*

Boeing 737-800, Irish Registration EI-CSC [4]

On 7 October 2000, the master caution light illuminated with the indication "OVERHEAD" on the main instrument panel as the aircraft was climbing and passing FL100. A second master caution light illuminated as the aircraft continued to climb, passing FL 127, this time with the indication that the passenger masks had deployed in the cabin. The Cabin Altitude Warning horn did not sound, as would have been expected. The crew requested and received clearance to descend back down to FL100. Since the aircraft was, by this time, less than 1.500 ft above the 10.000 ft target altitude, the Captain advised the First Officer that *"Theoretically at this point we should both be on oxygen but as we have only 1.500 ft to go we will skip it."* No emergency was declared. The senior cabin attendant was called to the flight deck to confirm that the oxygen masks had deployed in the passenger cabin, and was told that the aircraft would be returning to the departure airport because of a pressurization problem. The Captain made the relevant PA. Prior to commencing the descent, the crew analyzed the situation and discovered the Cabin Altitude horn circuit breakers in the "Open" position, and the two engine bleed air switches still in their preflight position (i.e., "OFF"). Restoring them to the correct position for flight returned the aircraft pressurization system to its normal condition. The aircraft continued its descent to the departure airport where it landed without further incident.

In its Final Report, the AAIU discussed standard procedures regarding the position of the engine bleed switches and noted that, per the normal checklist, they should both be "ON" before takeoff and remain "ON" for the duration of the flight. The AAIU also notes that SOPs provided the crew four separate occasions to evaluate/confirm the status of the pressurization/air conditioning system. Based on the events, the Unit concluded that *"It is clear that both flight crew members did not adhere to the Operator's SOP's with regard to completion and verification of the normal checklist. The PF failed to correctly configure the aircraft pressurization and air conditioning for flight. The PNF failed to visually confirm that the checklist responses were in agreement with the actual switch position. Neither flight crew member confirmed that the pressurization was functioning correctly."* The AAIU further stated that *"General airmanship should also provide for the monitoring of all aircraft instrumentation and on board systems during flight"* noting that *"... it is necessary to confirm from time to time the on-going integrity of the pressurization..."*.

With regard to the obvious failure of the Cabin Altitude warning horn to sound, the Unit found that the relevant circuit breaker was in the "Open" position, thus preventing the horn from sounding in the cockpit (the FDR record showed activation of the horn). The AAIU concluded that *"... failure to identify this condition* (correct configuration of circuit breaker switches) *during the pre-flight cockpit safety inspection denied the flight crew of a vital safety element of the pressurization system."*

Concerned about the *"... varying nature and effects of altitude/hypoxia on different individuals"* the AAIU suggested that *"... it would be considered prudent for flight crews to be required to don their oxygen masks:*

    i.   Immediately after a cabin altitude warning system activates; and/or

    ii.   When the cabin altitude exceeds 10 000 ft; and/or

    iii. When the integrity of the pressurization system is considered suspect."

The AAIU, finally, also discussed the role of the cabin crew in communicating non-normal situations to the flight crew and observed that *"Reporting on a significant change in the cabin environment could provide the flight crew with a vital source of information, which might assist them during their analysis of an emergency situation or a system failure."*

Boeing 737-700 Incident in Norway on 15 February 2001

On 15 February 2001, a Boeing 737-700 was climbing above 10.000 ft when the flight crew heard an aural warning horn, which the pilots associated with a takeoff configuration warning. Consequently, the flight crew determined it to be a false warning, and pulled the aural warning circuit breaker to silence the warning horn. Neither pilot associated the warning horn with a Cabin Altitude Warning. After approximately two minutes of level flight at FL250, the warning light indicating passenger oxygen masks deployment in the passenger cabin illuminated on the Overhead Panel. The flight crew immediately donned their oxygen masks and initiated a descent to 10.000 ft. It appeared that during the descent, the Air Conditioning Pack Switches were turned to ON.

The Norwegian Accident Investigation Board (AIB-N) investigated the incident and identified a number of active and latent failures, and promulgated five safety recommendations. One of the safety recommendations (no. 28/2002) called for the operator to evaluate, together with the Norwegian Civil Aviation Authority and Boeing, the dual use of the warning horn for Takeoff Configuration Warning and Cabin Altitude Warning, as well as the absence of a warning light for low cabin pressurization (over 10.000 ft) (as in use in MD-80 aircraft). The report on the incident by the AIB-N was issued on 17 July 2002.

NASA Aviation Safety Reporting System

The NASA Aviation Safety Reporting System (ASRS) conducted a search of its database for incidents similar to the accident flight. The result of the search yielded 171 reports of air conditioning and pressurization problems that involved Boeing 737 aircraft in the past decade (1994-2004). Of the 171 reports, 58 were deemed critical. According to Boeing, the rate of reports per a million departures was 2.7. (The number of reports for other aircraft was 94 on the B727, 97 on the B757/767, and 205 on the DC9/MD80. According to Boeing, the rates of reports per a million departures for these aircraft types were 8.5, 4.9 and 6.3, respectively).

The aviation industry formed a Pressurization Working Group (PWG). Its December 2005 report (No. 525) showed that the ASRS database contained 158 reports describing pressurization problems experienced by crews of Boeing 737 aircraft from January 1994 to October 2005. Slightly more than half of these involved aircraft of the same type as the accident aircraft, i.e. the -300 series.

The reports contained in both the above data sets described a large variety of aircraft pressurization failures or rapid decompression events experienced by flight crews. The reported causes ranged from the crew's failure to re-establish proper air conditioning and pressurization configuration after bleeds-off takeoff or single-engine taxi, to equipment malfunctions (e.g. doors not properly sealed and pressurization controller failure, outflow valve failure), to incorrect maintenance action (e.g. incorrect valve replacement), to non-adherence to published procedures, company policies, or FARs. It was interesting to note that a number of crews allowed the aircraft to continue climbing even after having clear indication of problems with proper pressurization of the aircraft.

Examination of the incident reports in the above two data sets yielded 10 incident reports of special relevance to the accident under investigation. These were reports in which the flight crews admitted to having, at least momentarily, misinterpreted a cabin altitude warning horn in flight to signify an improper aircraft configuration (i.e. takeoff configuration warning horn) rather than a pressurization problem. In only two such reports, the crew claimed having been predisposed to immediately think about a problem with the air-ground sensor upon hearing the warning horn because of another problem they had just been experiencing that was related to that sensor (auto throttle trip-off). In the remaining eight reports, the crew attributed their momentary confusion to subconscious predisposition and expectations, the result of training and exposure to line flying (e.g. simulator training typically presents decompression events during the cruise phase, and during line flying crews typically experience the horn less often, if ever, as a cabin altitude horn in the air than as a takeoff configuration warning on the ground).

Of interest and relevance to the accident under investigation were another nine reports of pressurization problems directly attributed to the crews' failure to set and verify the proper position of the pressurization mode selector to AUTO (Seven of these concerned Boeing 737 aircraft, while the other two events concerned McDonnell-Douglas aircraft). These nine reports all referred to aircraft that took off with the pressurization selector in inadvertently set to MAN (manual).

Three events were attributed by the reporting pilot to maintenance actions on the ground that had failed to return the pressurization mode selector to AUTO position – subsequent failure to notice the improper setting of the mode selector was attributed to conducting preflight checks in early daytime conditions with the cockpit light selector on dim, and fatigue and pressure to stay on schedule. In five events, the crew reported having failed to properly execute the preflight checks and consequently to verify the proper setting of the mode selector but could not explain why, except to speculate that their flow was interrupted, or that they thought they saw the selector in the expected position, or in a break in habitual action patterns due to unfamiliarity with the newer, digital pressurization system. In the last of the nine events, the mode selector was inadvertently left in the wrong position after an evacuation procedure demonstration to the fire department. The crew reported that the green light that illuminated when the mode selector was pointing to the MAN (manual) position was not discernible on the overhead panel in bright daylight conditions.

On 14 December 2004, the Director of NASA ASRS issued an Alert Bulletin entitled **"Boeing 737-300 Cabin Altitude Warning Horn",** addressed to the Boeing Company and copied for information to numerous FAA offices and international aviation industry organizations. Excerpts from the Bulletin included:

"A B737-300 flight crew reported to ASRS that they experienced a confusing and potentially misleading aural warning while troubleshooting a suspected air/ground sensing problem in cruise at FL370. Following disengagement of the auto throttle and failure of a transponder, the flight crew heard an intermittent warning horn sounding in the cockpit that they interpreted to be the takeoff warning horn. The first officer remembered that the takeoff warning horn also doubles as the cabin altitude warning horn. They realized the cabin was slowly losing pressure, donned their oxygen masks, and initiated a rapid descent to 10.000 feet .... The flight diverted to an alternate airport and landed."

In the narrative of the pilot's statement, he reported, in part:

"A safety issue I would like to raise awareness about based on my experience is the lack of wisdom in having the TKOF warning horn double as the ALT warning horn. Because the cabin was losing pressure slowly, we did not feel any pressure changes in our ears that would have normally served to alert us to a pressurization prob. If the FO had not happened to remember that the horn also serves as a cabin ALT warning horn, we may have continued trying to troubleshoot the air/ground prob, until passing out from lack of oxygen... Flt crews are not accustomed to associating an intermittent horn with the cabin ALT warning system. During the Capt's cockpit setup, the intermittent horn is tested by advancing the throttle to check the TKOF warning system. Like Pavlov's dogs, this forms the habit pattern of only associating that sound with that system (ALT warning) and as a result, there is no particular sound Flt crews are trained to associate with that system (TKOF warning)."

Other Incident Databases

The Canadian Transportation Safety Board and the Australian Air Transport Safety Bureau also reported occurrences involving pressurization problems that included takeoffs and climbs initiated with pressurization systems not properly configured, such as not opening the bleeds after a "no bleeds" takeoff, or not turning on the air conditioning packs for takeoff. One such incident on an aircraft led its Norwegian operator to forward a recommendation directly to the Manufacturer.

Previous Accidents involving Pressurization

LearJet 35 in Aberdeen, South Dakota, USA

On 25 October 1999, a Learjet Model 35 (registration N47BA) operated by Sunjet Aviation, Inc., took off from Orlando, Florida for Dallas, Texas. Eight minutes later, after the aircraft had climbed through FL230, the air traffic controller cleared the aircraft to FL 390 and the crew acknowledged the instruction. This was the last transmission from the aircraft. From this time and until the controller tried, unsuccessfully, to contact the crew later, the CVR contained the sound of the cabin altitude aural warning but no other sounds of conversation between the flight crew.

Based on this information, it was determined that the aircraft lost pressurization and that the cabin altitude reached a level at which consciousness could only be maintained with supplemental oxygen and which the flight crew failed to receive for unknown reasons.

The aircraft was later intercepted by several military aircraft as it tracked a northwest flight path. The military pilots flew at close enough range to determine there was no structural damage to the Learjet. However, they were unable to see in the cabin because the forward windshields of the Learjet seemed to be frosted or covered with condensation. At 12:13 local time (CDT) the aircraft impacted the ground in an open field.

The investigation determined that the *"probable cause of this accident was incapacitation of the flight crewmembers as a result of their failure to receive supplemental oxygen following a loss of cabin pressurization, for undetermined reasons."*

Beech King Air 200 in Burketown, Australia

On 4 September 2000, a Beech Super King Air 200 aircraft (registration VH-SKC), departed Perth, Western Australia for Leonora. For 23 minutes after takeoff, all communications with the pilot appeared normal. However, when the aircraft crossed FL250, the pilot's speech gradually became impaired until he deteriorated towards unconsciousness and unresponsive to air traffic control instructions. No sounds of pilot or passenger activity in the aircraft were detected for the remainder of the flight. About 5 hours after departing Perth, the aircraft impacted the

136

ground.

The investigation concluded that the aircraft occupants' *"incapacitation was probably a result of hypobaric hypoxia due to the aircraft being fully or partially unpressurized and their not receiving supplemental oxygen"* for undetermined reasons

[1]AAIU Report no. 2001-014, AAIU File No. 2000/2001, published 21 September 2001.
[2]AAIU Formal Report No: 2003/010, AAIU File No: 2002/0050, published 6 August 2003
[3]AAIU Formal Synoptic Report No. 2005-009, File No. 2004/0060, published on 23 May 2005
[4] AAIU Formal Synoptic Report No. 2001/0018, File No. 2000/0061, published on 30 November 2001

CHAPTER 10

ANALYSIS

General

The Captain and First Officer were licensed and qualified in accordance with applicable regulations and Operator requirements. Their duty time, flight time, rest time, and duty activity patterns were according to regulations. The cabin attendants were trained and qualified to perform their duties in accordance with existing requirements.

The air traffic controllers, who handled flight HCY522 were qualified for their duties and their performance was not a factor in the circumstances of the accident.

Visual meteorological conditions prevailed on departure and along the route of flight, and weather was not a factor in the accident.

The aircraft was properly certificated and equipped and maintained in accordance with regulations and approved procedures. The aircraft did not have significant pre-existing airframe or power plant problems. It departed with load data within limits and no deferred defect items were pending in the logbook. Nevertheless, the NVM from the cabin pressurization controller showed a pressurization leak or insufficient inflow of air of undetermined nature over an extended time period, which is discussed later in this report.

There was no evidence of an in-flight fire.

The Board examined the evidence to determine the causes of the accident and the analysis included examination of safety issues and safety deficiencies related to:

- Maintenance procedures and actions by the mechanics;
- Pilot training, procedures, and actions by the pilots;
- Organizational factors related to the Helios Airways flight operations and safety management systems;
- Organizational factors related to safety oversight of maintenance and flight operations by the Cyprus DCA, EASA, ICAO, and JAA;
- Organizational factors related to the airframe manufacturer's presentation of maintenance and flight crew procedures and management of precursor incident information to amend such procedures in a timely manner; and
- Organizational factors related to the aviation industry's management of precursor incident information to develop accident prevention measures in a timely manner.

As the result of the investigation, the Board identified numerous safety deficiencies which were communicated to the parties involved, some of which have already been resolved by safety actions taken by the responsible organizations. Additional safety deficiencies are addressed in the Board's report as safety recommendations for further actions to prevent future accidents

Sequence of Events

Pre-Departure Unscheduled Maintenance

The Board reviewed the maintenance actions performed on the aircraft by the Ground Engineers prior to its departure from Larnaca airport, as well as the Aircraft Technical Log entries and the AMM procedures/tasks recorded in the log book. A number of issues were of concern to the Board. The first set of issues was about the maintenance actions performed on the aircraft, both those recorded and those reported; the second set of issues was about the condition in which the aircraft was left for the flight crew after those maintenance actions, particularly with regard to the pressurization system.

Regarding the maintenance actions said to have been performed on the aircraft prior to its departure from Larnaca in the morning of 14 August 2005, the Board noted that Ground Engineer number one failed to correctly and completely document his work on the aircraft in the log book. With respect to the Cabin Pressure Leak Test, the Ground Engineer simply recorded *"Pressure run carried out to max diff"* without making reference to the procedure/task number followed for this *"pressure run."* According to the relevant AMM, Task 05-51-91-702-001, (entitled "Cabin Pressure Leak Test"), this test involved increasing the pressure in the aircraft *"...until the differential pressure indicator shows a differential pressure of 4.0 psi."* At that time, the outflow valve was to be fully closed and, with the help of a stopwatch, the decrease in cabin differential pressure from 4.0 to 2.5 psi was to be plotted. To determine whether the leakage rate was acceptable, the plot was to be corrected using a correction factor curve which was provided in the same task description. Of particular importance is the fact that paragraph D (7) contained a note stating that *"The indication on the differential indicator must not be more than 4.0 psi during a normal test"* when increasing cabin pressure.

According to statements by Ground Engineer number one, the cabin differential pressure during the test reached 8.25 psi. Therefore the differential pressure exceeded prescribed 4.0 psi and the absolute pressure in the aircraft exceeded 20 psi. This was the reason why he had to perform a flight crew oxygen mask test.

The Board noted that Ground Engineer number one operated both air conditioning packs for the leak test, instead of one pack, as prescribed by the AMM Task paragraph 6, which also contained a note that either the left or the right pack could be used. The Board was also concerned that the Ground Engineer number one did not appear to take notice of the warning in the AMM Task paragraph 2 which cautioned against *"... sudden pressure changes (that) will cause pain and injury and must not be done. When you do not obey the precautions, injury to persons will occur."* Furthermore, the Board noted that there was no mentioning of Ground Engineer number three having a printout of the AMM Task at his disposal in order to indicate to him which areas to check for possible leakage. The AMM Task listed areas to be checked for leakage from the outside of the aircraft when performing a Pressure Leak Test.

In his comments to the Board on the draft Final Report on the accident, Ground Engineer number one clarified the log entries as follows:

- As the numeric reference to *"8.25 ambient psi"* came from AMM 21-32-21-725- 001, he considered it appropriate as good practice to give the source document in his Log entry after the words *'No leaks or abnormal noises";*

- As a result, the whole line *"Safety valve operates at 8.25 ambient psi. No leaks or abnormal noises"* related to *"IAW AMM 21-32-21-725-001";*

- The previous line in the Log entry *'Door and local area inspected. Nil defects. Pressure run carried out to max diff"* related to AMM 05-51-91-702-001; and

- He did not include reference to 05-51-91-702-001 because of space constraints in the Log.

The Board, however, believed that space in the log book should not have been a factor, as there was always the option to continue recording on the next page of the log book. In fact, the note *"... USE NEXT PAGE"* appeared in the bottom right corner of each page. This option was actually exercised by the same engineer when he continued annotating onto the next page of the Log book to transcribe the remaining routine maintenance actions he also conducted on that same morning.

Another issue examined by the Board was that the AMM Task number recorded in the Technical Logbook by Ground Engineer number one, 21-32-21-725-001 entitled *"Test Safety Relief Valve Using Hamilton Standard Test Fixture"*, required the use of a Hamilton Standard Test Fixture. Ground Engineer number one stated that he had no test equipment with him when he performed the maintenance work. Furthermore, the above AMM Task was not referenced in the AMM Task that Ground Engineer number one reported to have been performing (Cabin Pressure Leak Test).

Ground Engineer number one stated that he referenced the AMM Task (21-32-21-725- 001 because he consulted that Task after conducting the maintenance work on the aircraft and returning to the office, to determine the value at which the safety valve operated – which he then recorded in the logbook (*"Safety valve operates at 8.25 Δpsi."*)

The Board wondered about the above described inconsistencies between the maintenance action reported to have been performed and the annotations subsequently made in the Aircraft Technical Log. The maintenance actions performed and described by Ground Engineer number one, as well as the pressure differential value attained, was not within the range of values prescribed by the intended AMM Task (05-51-91-702-001, Cabin Pressure Leak Test). The other AMM Task (21-32-21-725-001, Test Safety Relief Valve) was not required, nor possible in view of the absence of the required test equipment. The Board also wondered how it was altogether possible to raise the cabin pressure differential to 8.25 psi (as recorded by Ground Engineer number one in the log) using the APU because the APU supply alone is insufficient to raise it to this value and within the timeframe stated. The Board, finally, noted that Ground Engineer number one did not remain or return on board the aircraft when the flight crew arrived to prepare for the flight, to brief them about the work he had performed concerning the pressurization leak check.

Next, the Board considered the condition in which the aircraft was left for the flight crew after the maintenance actions. Specifically, the Board noted the fact that the pressurization mode selector had been switched to MAN (manual) position in the course of the Cabin Pressure Leak Test.

The condition of the pressurization system on the accident aircraft was evaluated based on the NVM data analysis at Nord-Micro. The NVM contains data from 42 records, which is the maximum capacity of the NVM. The oldest flight-leg-count read 8910 (internal CTR flight-leg-reading) and the latest read 8984 (the accident flight-leg) which represents a total of 74 flight-legs.

All these recordings indicated that the Digital Cabin Pressure Control System (DCPCS) was in the MAN (manual) mode, the Outflow Valve was constantly at 14,6 degrees opening angle (this was also the position the Outflow Valve Actuator was found at the accident site) and the flight mode was CLIMB. Cruise Flight Level was selected to 340 and Landing Field Elevation to 350 ft. The Cargo Heat Valve (Forward Outflow Valve) and both Pack Valves were indicated NOT CLOSED.

As far as the previous flight legs i.e., before the accident, are concerned, in flight leg 8973 a "22CAS 1-Fail" message was recorded. This indicated that the CTR (ConTrolleR) was not receiving the parameter CAS (Calibrated Air Speed) from DADC1 (Digital Air Data Computer No.1). This message was recorded on ground and does not represent a DCPCS problem or malfunction. Flight leg 8632 shows a message *"254 I BIT Entry"* which signals the initiation of a DCPCS calibration test on ground.

All other messages are *"030 INFLOW/LEAKAGE FAIL"* which are recorded when the Outflow Valve is below 3 degrees opening angle for more than five consecutive seconds and therefore barely able to keep the cabin pressure on schedule. All these messages were recorded immediately after take-off and they have a number of occurrences during the same flight as shown by the intermittent count of four, i.e. there are four or more events. Both Pack-Valves are indicated open as well as the Cargo Heat Valve.

All these messages indicated a continuous leakage situation with the aircraft that likely persisted for the last 74 flight legs. Only on the last flight is this situation not recorded, because the Outflow Valve did not modulate to compensate for the outflow of air, since the aircraft's Cabin Pressure Controller Mode Selector had be left in the MAN (manual) position. Based on the above indubitable evidence, the Board concluded that the pressurization mode selector was in the MAN (manual) position from the time the aircraft was still on the ground. It could not be determined whether the position of the selector was deliberately set to the MAN (manual) position sometime during the flight crew preflight duties.

However, the Board noted that the flight crew would have had no reason to deliberately set the pressurization controller to operate in the manual mode. Also, the Board noted that the flight crew would have had no reason to set the pressurization system to operate in the manual mode. In order to do so, there must have been a write up in the Log book or in the MEL guidance to use the manual mode due to an existing problem in AUTO or ALT. Furthermore, there was no evidence recorded of either pilot manually controlling pressurization on the aircraft during climb (i.e., by using the outflow valve toggle switch, as would have been the case if the MAN (manual) selection had been deliberate). Thus, the Board was led to believe that the selector had remained in the MAN (manual) position after the Pressure Leak Test, the last known time the particular selector had been manipulated. Naturally, the fact that the mode selector position was not rectified by the flight crew during the aircraft preflight preparations was crucial in the sequence of events that led to the accident, and the Board also considered this scenario in depth.

Another indubitable piece of evidence adding to the above was the unusual, excessive rate of climb of cabin altitude. A calculation of the rate of climb of cabin altitude during the initial stages of climb show the average cabin rate of climb to have been about 2.000 ft / min (the cabin climbed 10.000 ft in a little more than 5 minutes when the aircraft reached 12.500 ft). This rate exceeded by far the rate of climb normally achieved when pressurization system was in the AUTO mode. The Board, therefore, concluded that the aircraft pressurization system could not have been in AUTO mode when the aircraft departed the ground. Therefore, there was no automatic control of the OFV to regulate the cabin rate of climb. The OFV remained fixed (open) in the position it was last left in. Thus, the cabin pressure followed quite closely the aircraft rate of climb. This would explain the excessive rate of climb of cabin altitude.

Finally, the Board considered it important to determine whether the AMM Task, "Cabin Pressure Leak Test", called for the pressurization mode selector to be returned to the AUTO position following such a test. AMM Task 05-51-91-702-001 was examined. Its last action item stated *"F. Put the Airplane Back to its Initial Condition"* and specifically contained 3 actions sub-items as follows:

1. checking the pitot static system, if necessary,

2. opening the equipment cooling flow control valve, and

3. returning the equipment cooling fan switches to NORMAL if previously selected to ALTERNATE.

The Board was of the opinion that the AMM instructions were vague in making such a broad reference to an aircraft's *"initial condition"* and that the three actions items listed did not include returning the pressurization mode selector to the AUTO position, although the AMM Task explicitly required setting that selector to the MAN (manual) position for the Test. In this respect, the Board concluded that if the Ground Engineer number one had not returned the mode selector to the AUTO position, this could not be considered an omission as there was no specific requirement to do so. However, although the Board further noted that there was a requirement for the flight crew to ascertain that all selectors, including the pressurization mode selector, were in their proper positions for flight during their preflight preparations, the Board believed that it would have been prudent for the Ground Engineer number one to verify return of the pressurization mode selector to the AUTO position.

Preflight

The Board considered various aspects of the flight crew's performance, beginning with the preflight phase and until their last known communication. All available data from the FDR and statements from air traffic controllers and company pilots to gain insight into possible actions by the flight crew were used. The lack of precise records (e.g. a voice transcript) of flight crew communications for that time period meant that the Board had to rely on their operational and human factors expertise to provide a plausible account and an educated interpretation of the activities on the flight deck.

Following FCOM procedures, in the morning of the accident, the flight crew was expected to prepare the aircraft by first accomplishing their respective preflight duties that included checking the Equipment Cooling switches, the Cabin Pressurization Panel, and the flight crew oxygen masks. As this was the first flight of the day and the crew was starting the duty cycle, the pilots would have been expected to be particularly alert in conducting their preflight duties (Preflight Procedure, Before Start Procedure, Before Taxi Procedure, and Before Start Checklist).

However, there was evidence that the flight crew had not performed their preflight duties as outlined in the relevant sections in the FCOM (see 1.17.2.2) in their entirety and/or correctly. When the aircraft departed, the pressurization mode selector remained in the MAN (manual) position (instead of AUTO) and remained there until the aircraft impacted the ground almost three hours later.

Both the Captain and the First Officer were experienced pilots and had performed the preflight duties numerous times in the past. The Board examined the reasons why they could have made such a crucial omission. In general, when the pressurization mode selector is positioned to the MAN (manual) position, it is accompanied by an advisory, green light indicating MANUAL. Normally, with the mode selector on AUTO as prescribed by the Preflight Procedure, no illuminated indication should appear on the pressurization panel. Given that the aircraft departed the gate area with the selector in the MAN (manual) position, the Board examined why an experienced crew would have failed to notice the presence of an indication they did not normally expect to see at this location in this phase of flight. In fact, it was an indication that flight crews did not normally see in any phase of flight, unless the selector was deliberately moved to the MAN (manual) position, i.e. if the crew was prepared to intentionally fly the aircraft using the manual pressurization mode because, for example, of a failure of the automatic pressurization system.

The Board considered various factors that may independently, or in conjunction, explain the failure to properly configure the pressurization panel during preflight. One factor concerned the visibility of the illuminated indication (MANUAL) given ambient light conditions at preflight. During the flight re-enactment, the Board compared the perceptibility of flight deck annunciations in both the BRIGHT and DIM settings of the lights switch. With the lights on DIM and looking up to the overhead panel from the First Officer's seat, the green annunciation MANUAL was not particularly obvious. Experienced pilots would typically position the lights switch to the setting that was most appropriate for ambient conditions at the time. Therefore, the Board believed that the flight crew conducted their preflight duties with the lights switch set to BRIGHT since they conducted their checks at daylight, between 08:30-09:00 h local time for a 09:07 h local time takeoff (sunrise at 06:05 h local time). The wreckage findings also revealed that the switch was set to BRIGHT at the time of impact.

The Board believed that the preflight duties were most likely conducted with the lights set to BRIGHT, and that the green indication should have been clearly visible. The Board examined why the two experienced pilots failed to note its presence. From a human performance standpoint, an individual may not notice the presence of a cue when he or she does not expect that cue because of a natural human vulnerability to "look without seeing" when performing lengthy, habitual activities. A typical Preflight Procedure may contain between 40 and 80 actions to be performed by the First Officer, often under the pressure of the impending departure, and in the presence of a Captain who is waiting to call for the ensuing checklist. This procedure is performed from memory, aided by the fact that the actions are organized along the topographical location of panels in the cockpit. Memorization is beneficial for long lists of actions, but has the disadvantage that actions are performed automatically, without conscious effort and attention. This can and has, in the past, led to inadvertent omissions and other types of mistakes.

The Board was also sensitive to the fact that automatic execution of actions was very much affected by assumptions – in the case of performing a large number of verification steps, the assumption that all switches and indications were in the usual, normal for this phase of flight position. A superfluous green indication on the pressurization panel could be easily (inadvertently) overlooked when perception was biased by the expectation that it should not be present.

Exacerbating this tendency (expectation bias) is the rarity with which switches (especially, and directly relevant to this case, the pressurization mode selector) are in other-than-their-normal position. A pilot automatically performing lengthy verification steps, such as those during preflight, is vulnerable to inadvertently falsely verifying the position of a switch to its expected, usual position (i.e. the pressurization mode selector to the expected AUTO position) – especially when the mode selector is rarely positioned to settings other than AUTO.

Insofar as the preflight duties were concerned, the Board concluded that the crew's omission of a critical check, while serious, could be partially explained by human performance factors that have affected procedural execution in previous accidents. The Board was concerned that the overhead panel design was not conducive to safeguarding against these types of inadvertent omissions. Specifically, the color of the illuminated indication (green) does not typically imply something out of the ordinary, as did the amber (caution) or red (warning) – which would have likely attracted the flight crew's attention that something was out of the ordinary.

After the Preflight Procedure, the crew was expected to orally execute a Preflight Checklist. Per the carrier's FCOM, this checklist included a check of the pressurization panel:

(item 12 of 25) AIR COND & PRESS ___PACK(S), BLEEDS ON, SET

The flight crew failed to detect the improper configuration of the pressurization panel during this checklist. Both the Captain and First Officer had repeatedly accomplished this checklist on many flights during their long careers. Their failure to properly accomplish the above checklist prevented them from capturing their earlier mistake. This was the first of two missed opportunities to notice and correct an earlier error. Various factors could have contributed to this failure.

The Board first examined the design of the checklist, and specifically the fact that the challenge part of this action item *("Air Cond & Press")* essentially combined two separate systems (air conditioning and pressurization). While this combination was certainly not random (the two systems used engine bleed air as an energy source), the corresponding response portion of the action item *("Pack(s), Bleeds ON, SET")* contained three different confirmations, only the third of which referred to the pressurization panel. In turn, this third confirmation referred to eight different actions – those that were performed earlier, during the Preflight Procedure. Contrary to the manufacturer's original intention, however, many pilots informally reported that when performing the checklist and responding *"SET"* to the pressurization panel, they really only checked that the landing and cruise altitudes had been correctly set in the corresponding indicators.

The performance of checklists in routine, daily flight operations was also examined. In general, checklist items are performed by referencing a printed card. Like procedures, because they are performed repeatedly on the line, they are also performed by memory, typically in time-pressured circumstances (i.e. indirect pressure to maintain on-time departures). For these two reasons, checklists are often performed in a hurried, automatic fashion. From a human factors standpoint, rushing is known to lead to the inadequate allocation of attention to the task at hand – and thus to errors. Furthermore, like procedures, checklists are also vulnerable to "looking without seeing" because they are biased by the assumption that since each item verified an action performed only moments ago, then it must be already in the desired position/set.

In this respect, the Board noted that the presence of a number of relevant comments in the First Officer's training record. These comments were written by training and check pilots who, in the course of simulator and line flights, had observed the First Officer to have difficulty with following SOP and with the correct usage of checklists (mistakes, omissions), and to display a tendency to overreact and lose confidence in non-normal situations. The Board believed that the type of performance observed in training could have been a factor in this first, critical omission of verification of a checklist item (i.e. the pressurization mode selector).

Later on, following engine start, the crew prepared the aircraft for taxi. In performing the Before Taxi procedure per the FCOM, pilots were expected to verify that all system annunciator panel lights illuminate (amber) and then extinguish (i.e. a "recall" check). This check was performed by pressing on either of the two annunciator panels. By design, nothing on the two annunciator panels, however, would have provided any clues to the flight crew regarding the improper setting of the pressurization mode selector. The Board was concerned that if the MAN (manual) position of the mode selector had been designed to be accompanied by an annunciation upon recall (e.g. AIR COND), the flight crew's attention would have been directed to the corresponding panel and the error may have been rectified.

As analyzed above, the selector was in the MAN (manual) position from the beginning of the flight and that the green indication MANUAL was already illuminated during the preflight checks.

Takeoff

Following takeoff, the flight crew was to perform an After Takeoff checklist, the first item of which was to check the pressurization system again and verify its settings. Although this checklist would have directed the flight crew's attention to the pressurization panel, there was no evidence that the incorrect position of the pressurization mode selector was rectified. This was the second missed opportunity to note and correct an earlier error. It was not possible to determine why the flight crew failed to notice the incorrect setting of the pressurization mode selector, but the human performance issues mentioned above, including the documented issues in the First Officer's training record, were also likely to have affected their actions at this time. In fact, the After Takeoff checklist is also usually performed under even more time-pressured conditions and at a time when the pilots' attention is consumed by other, concurrent tasks (e.g. retracting the landing gear and flaps, monitoring the climb, and communicating with ATC). The management of multiple concurrent tasks requires the division of attention resources and is known to force a person to devote insufficient attention to any one of the many tasks.

Climb

At an aircraft altitude of about 12.000 ft the cabin altitude warning horn sounded. Eight seconds later, the FDR showed the autopilot being disengaged, and re-engaged four seconds later. Eight seconds later, the FDR showed the auto-throttle being disengaged and the throttles retarded, but like the autopilot it also was re-engaged nine seconds later. Three seconds later, the No.2 radio was used to contact the Helios Airways Dispatcher.

The Board examined the flight crew's actions to disengage the autopilot and auto-throttle, and to retard the throttles upon onset of the warning horn. Given that the expected reaction to a cabin altitude warning horn would have been to stop the climb (there was no evidence to this effect), the Board considered such actions to signify that the flight crew reacted to the warning horn as if it had been a Takeoff Configuration Warning (the two failures use the same warning horn sound). This possibility was strengthened by the Helios Dispatcher who reported that when the Captain initially contacted him on the company frequency, he had referred to a takeoff configuration warning horn. This statement could not be verified because no voice transcript of this communication existed. Had the flight crew realized the significance of the warning horn, they should have immediately donned their oxygen masks, per the Operator's Standard Operating Procedures.

The Board concluded that the flight crew confused the two meanings of the warning horn. Similar occurrences had been reported by flight crews worldwide in the past. Given the benefit of hindsight, confusion of two meanings of one horn (one of which was designed to only sound when the aircraft was on the ground) appeared irrational. Various factors for creating the potential confusion of the two experienced pilots were considered.

The warning horn was designed to signal two distinctly different situations.

The Board considered the role of experience in interpreting and reacting to the warning horn. In the course of his career, a pilot is generally likely to only hear the warning horn when it is associated with a takeoff and a takeoff configuration problem. This occurs in one of two ways: when the crew mistakenly attempts to takeoff without having properly configured the aircraft for takeoff (i.e. set trim, flaps, and speed brake); and when the Captain momentarily advances the throttles during the preflight check or immediately before takeoff as a precaution to verify that the warning horn is operating. The second case does not require any immediate action since it is a deliberate action that is immediately canceled when the Captain pulls the throttles back to idle. The first case, however, demands that the pilot immediately retard the throttles and abort the takeoff. Both declarative memory (i.e. the type of memory that stores facts and events) and muscle memory (i.e. skeletal muscle activity that becomes essentially automatic with practice) associated with onset of this warning horn in the cockpit are directly and intricately linked to the throttles and hence the takeoff significance of the horn.

Most pilots are not very likely to experience a cabin pressurization problem and the associated warning horn at any time during their line flying.

The Board also considered the role of stress that probably further contributed to the possible confusion of the two meanings of the warning horn. In general, stress, such as that caused by the onset of a loud, distracting alarm in the cockpit, combined with the element of surprise, is known to lead to automatic reactions. Automatic reactions, in turn, are typically those that result from experience and frequency of encounter and are therefore not always appropriate. The Board considered that the flight crew may have automatically reverted to a reaction based on memory before consciously processing the source and significance of the stress factor. This would also explain why the flight crew failed to realize the improbability of their interpretation of the horn as a takeoff configuration horn and why they failed to move on to gathering information for a new, correct diagnosis of the problem at hand. To support the notion that stress interfered with what would now be considered a "normal" reaction to a warning, it was important to note that at no time during this sequence of events was the cabin altitude warning horn cancelled. The flight crew certainly had the option (and, in fact, was supposed) to push a button to cancel the horn in order to avoid the distraction it was causing while trouble-shooting the source of the problem.

The Board also entertained the possibility that the Captain really meant to say "*Cabin Altitude Warning horn*" when he contacted the Dispatcher at the Helios Operations Centre and that he was, in reality, aware of the difference – but referred to the warning by its most frequently encountered name (i.e. takeoff configuration horn). As English was not the Captain's native language and under the influence of stress, this possibility could not totally be ruled out. However, the actions taken by the Captain were indications against this assumption and supportive of an interpretation of the horn as a takeoff configuration warning.

In general, sufficient ease of use of English for the performance of duties in the course of a normal, routine flight does not necessarily imply that communication in the stress and time pressure of an abnormal situation is equally effective. The abnormal situation can potentially require words that are not part of the "normal" vocabulary (words and technical terms one used in a foreign tongue under normal circumstances), thus potentially leaving two pilots unable to express themselves clearly. Also, human performance, and particularly memory, is known to suffer from the effects of stress, thus implying that in a stressful situation the search and choice of words to express one's concern in a non-native language can be severely compromised.

According to FDR data, less than a minute into the radio communication exchanges between the Captain and the Helios Operations Dispatcher, at an aircraft altitude of about 17.000 ft, the MASTER CAUTION light was activated and was not canceled for 53 seconds. Two different events occurred at about this time, either one of which would have triggered the MASTER CAUTION light with the accompanying OVERHEAD indication on the Annunciator Panel to draw the attention of the pilots to a situation indicated on the Overhead Panel. The equipment cooling low flow detectors reacted to the decreased air density and one or both of the Equipment Cooling lights illuminated on the Overhead Panel. In addition, the oxygen masks deployed in the passenger cabin, illuminating the PASS OXY ON light, located further aft on the Overhead Panel. The Board was unable to determine which event occurred first and triggered the MASTER CAUTION. However, the fact that the flight crew had not cancelled the first MASTER CAUTION meant that the second event did not trigger a second MASTER CAUTION as it was already on. Consequently, there was nothing to prompt the flight crew to look for a second indication on the Overhead Panel.

The Helios Dispatcher reported that the Captain referred to the Takeoff Configuration horn, and then the Equipment Cooling lights. After that, he was unable to understand the Captain's concerns and suggested the Captain talk to the on-duty Ground Engineer. At this point, the flight crew may have also started having additional indications in the cockpit, such as a flickering or illumination of the Equipment Cooling normal and standby fans OFF lights. According to the Ground Engineer number one, who had performed maintenance work on the aircraft that morning, the Dispatcher did not inform him of what the Captain had already reported (i.e. Takeoff Configuration Warning and the Equipment Cooling lights) before placing him on the microphone. The Ground Engineer stated that he did not notice the sound of the warning horn in the background during his communications with the Captain. When Ground Engineer number one communicated with the Captain, the latter told him that both the Equipment Cooling lights were off. The Board believed the Captain was trying to say that the Equipment Cooling fan OFF indications were illuminated, indicating a problem with the operation of the fans. In his statement after the accident, Ground Engineer number one stated that he was not aware that these indications were labeled OFF – and so the Captain's statement did not immediately make sense to him. After the Captain asked him about the location of the associated circuit breakers, the Ground Engineer asked him to confirm that the pressurization mode selector was set to AUTO. In his statement, he said that he asked this question despite the fact that he had no doubt he had returned the selector to the AUTO position after the maintenance work which he had performed that morning. The Board concluded that the question to the Captain indicated that the Ground Engineer was concerned about the position of the mode selector.

The opinions collected by the Board generally differed on whether there were any difficulties in communication between the Captain and the Helios Operations Centre. In particular, according to some statements there were difficulties due to the fact that the Captain spoke with a German accent and could not be understood by the British engineer. The British engineer did not confirm this, but did claim that he was also unable to understand the nature of the problem that the Captain was encountering. The Operations Dispatcher then suggested that the Cypriot First Officer be asked to talk in Greek with the Cypriot Engineer also on duty. The language difficulties prolonged resolution of the problem, while the aircraft continued to climb. Moreover, the communication difficulties could also have been compounded by the onset of the initial effects of hypoxia.

Twenty seconds later, at an aircraft altitude of about 20.000 ft (FDR data), the MASTER CAUTION was canceled. During this entire time period, since its onset at 06:12:38 h, the cabin altitude warning horn continued to sound. The Captain continued to talk on the radio to the Helios Dispatcher. There was no other evidence, however, to suggest that either of the two pilots was aware at this point that the cabin altitude was rising since the aircraft pressurization was very poor and most of the inflow mass of air was outflowing (overboard) causing the aircraft cabin altitude to increase (in conjunction to aircraft altitude) with very limited differential pressure as a result.

The appropriate action would have been to start a descent, or at least level the aircraft. There was no evidence to this effect.

At the time of onset of the MASTER CAUTION (and the OVERHEAD indication), workload in the cockpit was already high. The crew was facing a loud warning horn and the realization that the attempted actions (to disconnect the autopilot and autothrottle and retard the thrust levers) had failed to address whatever situation was evolving. Their attention, furthermore, had already been diverted to the source of the initial Master Caution indication. All these factors may help explain why they did not notice the more serious indication of the deployment of the oxygen masks in the passenger cabin.

Given the ongoing distractions, the Captain, at least, may never have consciously and fully registered the onset of the indications and/or their significance. Unfortunately, although there were partial data to somewhat deduce the Captain's actions at this time (from his communication exchanges with the Operator's dispatcher and engineer), there was no possibility to establish the First Officer's actions during this same time. For example, he may have assumed that the Captain had the situation under control and may have therefore canceled the Master Caution before the Captain had an opportunity to consciously register its onset. It was also possible that, because the Captain was busy communicating with the Helios

Dispatcher, the First Officer silently pointed to the new indication he had observed (PASS OXY ON) and then canceled the warning thinking the Captain had acknowledged it. The Board was led to again consider the comments recorded in his training records and which indicated a tendency to overreact and lose confidence in non-normal situations. This observation, coupled with sub-optimal crew coordination would have undoubtedly reduced the potential for the First Officer to help the Captain manage the evolving non-normal situation. Of course, it was also possible that the First Officer was already beginning to become affected by the hypoxic conditions that were now subsuming the flight deck.

The Board was therefore concerned that the flight crew might have been subconsciously misled into believing the situation at hand was the result of a failure of the Equipment Cooling system. This would explain the on-duty Ground Engineer's claim that the Captain asked him for the location of the Equipment Cooling circuit breakers. (The Ground Engineer replied that they were behind the Captain's seat). The Captain may have left his seat at this moment in time in order to gain access to the equipment cooling fan circuit breakers.

The Board recognized that from a human factors standpoint, preoccupation with one task (i.e. trouble-shooting the source of the Equipment Cooling problem) at the expense of another (i.e. trouble-shooting the source of the warning horn) was entirely plausible and has happened to experienced pilots (e.g. preoccupation with a landing gear position indication was one of the primary factors in the 1972 accident that led an L-1011 aircraft into the Florida Everglades).

A number of factors related to preoccupation were considered. The possibility that the flight crew may have already been suffering from the initial effects of hypoxia must be taken into account. Gradual hypoxia symptoms included a degraded ability to think clearly, especially through novel situations, and degraded night/color vision. Although the Board was unable to explain how the flight crew may have failed to notice the physiological symptoms of the insufficient aircraft pressurization (primarily its effects on the ears) it could definitely ascertain that such symptoms affected pilots' decision-making and performance at some point during the climb. Another factor in cases of preoccupation was the presence of warning horns. In general, loud warning horns and alarms, especially when they are allowed to continue in time, can have the unintended side-effect of turning into strong distractions, rather than alerts. The loud cabin altitude warning horn that was allowed to continue to sound undoubtedly prevented the flight crew from clearly focusing on the situation at hand. The B737 was equipped with an ALT HORN CUTOUT button located between the cabin altitude indicator dial and the pressurization control panel. The Board noted that in order for the flight crew to silence the warning horn by pressing the cutout button, the pilots would have had to recognize that they were dealing with a pressurization problem.

The combination of hypoxia and distractions generally increases stress levels. Stress is known to render human cognition (e.g. memory, attention, decision-making, risk management, communication skills) particularly vulnerable to errors.

At an aircraft altitude of 28.900 ft (FDR data), six minutes after the first time the Captain contacted the Helios Operations Centre, the FDR showed the last keying of the No. 2 VHF radio. This marked the last known attempt of radio communication by the flight crew of HCY522. The VHF radio remained tuned to the Nicosia ACC frequency as was evident from the CVR which later (within its recording duration) recorded radio calls from other aircraft to the Nicosia ACC.

Pilot Incapacitation

The flight crew's last communication with Nicosia ACC was at 06:11:54 h, when the Captain acknowledged the clearance to FL340 direct to the RDS VOR. Until 06:11:54 h, his speech was normal and the phraseology was accurate.

After that time, no other communication with Nicosia ACC or any other ATC unit was recorded. The flight crew's failure to respond to repeated ACC radio inquiries from 06:30:40 h up to 06:34:44 h (six times) was the first indication of an existing problem.

Between 06:35:10 h and 06:39:30 h, Nicosia ACC called flight HCY522 five times without receiving a response. At 06:35:10 h, Nicosia ACC requested another aircraft (CYP498) to call HCY522. At 06:35:49 h, Nicosia ACC instructed flight HCY522 to squawk the STANDBY mode, and at 06:39:30 h, called the flight on the emergency frequency (121.5 MHz). There was no response from the accident aircraft to any of these calls. There were no further communication attempts from Nicosia ACC. As the flight continued, the aircraft did not deviate from its assigned route and flight level.

According to the FDR, the microphone keying (communication between the Captain and the Helios Operations Centre) ended at 06:20:2 1 h as HCY522 was passing through 28 900 ft. The sequence of events indicated that the flight crew became incapacitated at some point after 06:20:2 1 h.

The continuous sounding of the cabin altitude warning horn (FDR data) indicated that the aircraft and its occupants were in an environment that exceeded a cabin altitude of 10.000 ft and experienced progressive loss of cabin pressure during climb to FL340. The amount of oxygen in the cabin due to the inadequate pressurization became insufficient for the flight crew to maintain their consciousness.

Cruise

According to the statements submitted by the F-16 aircraft pilots that intercepted flight HCY522 during its cruise at FL340, the Captain's seat was vacant and the First Officer was slumped over the aircraft controls. The Board believed that the Captain got out of his seat possibly to check the circuit breakers. The flight crew oxygen masks were not found in the wreckage but the First Officer was observed later (when the aircraft descended) to not be wearing an oxygen mask. Therefore, the Board believed that the flight crew failed to don the oxygen masks and make use of the flight deck oxygen system. Thus, they succumbed to the effects of the hypoxic conditions. The Board believed that the DNA finding that the flight deck observer oxygen masks contained biological material that was consistent with the DNA of the First Officer was a result of the severe impact forces.

The F-16 pilot stated that the flight deck and cabin windows were free of mist and frost. He observed no detectable movement in the cabin by passengers or cabin crew. In fact, he only reported seeing three passengers seated and wearing oxygen masks. The fact that the passengers were seen wearing oxygen masks confirmed that the passenger oxygen masks deployed from the Passenger Service Units, as they were designed to do when the cabin altitude exceeded 14.000 ft, and as indicated by the FDR data. Furthermore, at least some of the passengers were able to make use of these masks and the Board concluded that at least some of the aircraft occupants were conscious during the climb to cruising level.

It was not possible to determine whether any of the cabin crew members were conscious during the climb, if they had been in a position to don oxygen masks as prescribed by the procedures, and whether they demonstrated or instructed the passengers in the use of the oxygen masks that had deployed. The amount of oxygen supplied by the passenger oxygen system was designed to last 12 minutes. In order to retain consciousness after the depletion of the oxygen from the passenger oxygen system, a person on board would have had to make use of one of the additional means of oxygen supply available on board the aircraft, i.e. the portable oxygen bottles. All four oxygen bottles were retrieved from the wreckage; three bottles were found with their valves in the open position. The Board concluded that these bottles were most likely used by someone on board the aircraft.

The observations by the F-16 pilot indicated that the aircraft never suffered any structural or mechanical damage, either prior to or during the intercept. There was no evidence of fire or smoke, or any fluids (hydraulic, fuel) from the aircraft – and hence no evidence of loss of aircraft control. The aircraft continued to fly along the FMS-programmed route via the EVENO and RODOS waypoints. During this time, the Athinai and Nicosia Air Traffic Controllers repeatedly called flight HCY522 and exhausted all available means to establish contact with the aircraft, but without success.

Descent

As the aircraft flew the KEA holding pattern at FL340 for the tenth time, at 08:48:05 h, the F-16 pilot reported seeing a male person wearing a light blue shirt and a dark vest, but not wearing an oxygen mask, enter the cockpit. This was confirmed by the CVR transcript. The sounds identified matched those of someone using the prescribed access procedure to enter the cockpit, followed by sounds similar to the flight deck door opening. The person proceeded to sit down in the Captain's seat. At the same time, the CVR transcript contained sounds that were identified as the inflation of an oxygen mask harness. The Board believed that the F-16 pilot may not have been able to observe an oxygen mask on the person's face, because the portable oxygen bottle mask was clear in color.

According to the FDR data, at 08:49:50 h, the left engine of the accident aircraft flamed out due to fuel starvation. This was confirmed by the statement of the F-16 pilot that fumes were observed to come out of the left engine exhaust pipe, which was a normal indication of engine flameout in flight. At this time, the aircraft exited the holding pattern by starting a left descending turn and followed an uneven flight path of fluctuating speeds and altitudes. The Board considered this evidence that the person in the Captain's seat was making an effort to control the aircraft. The F-16 pilot followed the accident aircraft, continuing to attempt to attract the person's attention but without success.

During the initial descent, at 08:54:18 h, the CVR record contained a MAYDAY call from the person in the Captain's seat. The call was not transmitted over the VHF radio; it was only picked up the CVR microphone. The second MAYDAY call was at 08:55:05 h followed by a third one a few seconds later. Based on the fact that there was only one male cabin attendant on board the accident aircraft, that the voice on the CVR was identified by colleagues to match that of the male cabin attendant, and that the person that entered the cockpit was wearing a Helios cabin attendant uniform, the Board concluded that the person that entered the cockpit and made efforts to control the aircraft was the male cabin crew member.

During this time, the aircraft continued to descend towards the ground. Only once did the person in the Captain's seat appear to notice the F-16 and responded to his hand signals, but there was no evidence that he attempted to follow the F-16 aircraft. The Board believed that any person with the cabin attendant's commercial pilot license background, under the prevailing conditions of potential hypoxia and extreme stress, would have been unable to gain control of a B737 with one engine stopped due to fuel starvation.

The second engine flamed out at 08:59:47 h, also due to fuel starvation. The aircraft continued to descend without engine power and without electrical power except for the instruments and systems which were powered by the aircraft battery. It was not likely, nor reasonable to assume, that the APU would have been started for electrical power. It impacted the ground at 09:03:32 h. According to the observations reported by the F-

16 pilot and the way in which the aircraft impacted the ground, the person at the controls appeared to have made an attempt to level the aircraft to alleviate the impact.

Cabin Crew Performance

No data existed to establish the cabin crew preflight activities with any certainty. It was assumed that before boarding the aircraft, the four cabin crew members participated in the required pre-flight safety briefing, and that the briefing was conducted together with the flight crew. The length and content of this brief, the manner in which it was administered, the time in which it was accomplished, the conditions under which it was carried out, and whether everyone was present, were all factors that likely set the tone for crew interactions during the accident flight. The Board could not determine whether the Captain invited the cabin crew to feel free to communicate with the flight deck during the flight if necessary (i.e. other than the required "safety checks" every 20 minutes) and if so, under what circumstances he was open to calls from the cabin crew. During post-accident interviews with other cabin crew members, the Board did not receive any reports that the Captain set up any type of negative climate wherein he did not allow the cabin crew to seek contact with the flight deck, if and when the need arose.

Data from the accident site indicated that the galley carts were stowed when the aircraft impacted the ground. The Board believed that the cabin crew had not yet started preparations for passenger service.

The first indication that the flight was not evolving as expected was probably noticed by the two cabin crew members seated in the jump seats next to the forward galley. Given their position right outside the flight deck door, it was possible that the Cabin Chief and another cabin crew member could have heard the loud Cabin Altitude Warning horn in the cockpit. Although it was not very likely that they would have known the significance of this loud auditory warning, it could not be determined beyond doubt that this was the case.

Two and a half minutes later, all four cabin crew as well as the passengers would have abruptly become aware of something out of the ordinary when the passenger oxygen masks deployed out of their units above the seats. Acting per procedure, the cabin crew would have reach out and donned the nearest oxygen mask. It was not possible to determine if the cabin crew acted per procedure, and if so, whether they all did, but post-accident interviews with other cabin crew members ascertained that all cabin crew were properly trained to execute this first step in reaction to the deployment of the oxygen masks. The Board had no reason to believe that this cabin crew would have done otherwise.

Similarly, it was not possible to determine whether the passengers followed the cabin crew's instructions (given during the preflight passenger safety briefing) or followed their example at that given moment. It was possible that some passengers were asleep, or slightly hypoxic already, or encountered problems in trying to don their oxygen masks. It could also not be determined whether any of the cabin crew members attempted to assist any of the passengers. This was not prescribed by the Operator's procedures which specifically instructed cabin crew to remain seated with their seat belts fastened. The statement by the F-16 pilot that he observed three passengers wearing oxygen masks but many masks dangling from their overhead units led the Board to believe that the oxygen masks were likely not worn by many passengers.

The Board considered whether and at what point the cabin crew may have experienced hypoxia symptoms. Based on a review of the Operator's manuals, cabin crew were probably sufficiently trained to recognize the physical characteristics of lack of pressurization. This, however, did not necessarily mean that the cabin crew actually recognized them or that they did so in a timely manner on this flight. Furthermore, the relevant guidance available in the Flight Safety Manual (FSM 4-70 to 4-7 1) mentioned the two types of decompression (slow and rapid/explosive) but only discussed expected symptoms of the latter type. All ensuing guidance concerned actions to take in the event of rapid or explosive decompression.

According to the procedures, the cabin crew members would have expected the aircraft to start descending or at least level off when the oxygen masks deployed in the passenger cabin. It was not possible to determine to what extent the cabin crew was cognizant or affected by the on-set of hypoxia, when the two indications that something was out of the ordinary occurred. Later, they would have expected an announcement from the flight deck calling the Cabin Chief to the cockpit and informing the passengers that it was safe to remove the oxygen masks. During this flight, however, the aircraft continued to climb.

The Board evaluated what the cabin crew's reactions might have been when the aircraft continued to climb and there was no announcement from the flight deck. There was no Operator procedure to address such a contingency. As emphasized by the Cabin Crew Manager in his post-accident statement, however, cabin crews were encouraged to take initiative. The Manager expressed his conviction that the particular cabin crew was well trained and by nature fully bound to have taken the initiative to seek an explanation for the unusual situation they were facing. The Board considered the fact that even if this was the case, it was hard for a cabin crew to assess how long to wait before contacting the flight deck – and in this case, time was of the essence as the hypoxia effects grew increasingly stronger. It was not possible to determine whether any of the cabin crew members attempted to contact the flight crew or enter the flight deck.

Data from the CVR only contained to the last 30 minutes of the accident flight and showed that at least one cabin crew member retained his consciousness for the duration of the flight and entered the flight deck more than two hours after takeoff. At the beginning of the climb phase, this cabin attendant was likely seated next to the aft galley. In order for him to have moved forward in the aircraft to reach the flight deck, he must have used a portable oxygen bottle.

The Board found the fact that this cabin attendant might not have attempted to enter the flight deck until hours after the first indication that the aircraft was experiencing a non-normal situation quite puzzling. Of course, in the absence of a longer-duration CVR, it was not possible to know whether this or any other cabin crew member had attempted to or succeeded in entering the flight deck. From the sounds recorded on the CVR, however, the Board could ascertain that this cabin attendant entered the cockpit using the emergency access code to open a locked cockpit door.

Crew Resource Management

The Board examined Crew Resource Management (CRM) issues related to the sequence of events involving flight of HCY522 in order to understand underlying reasons for the accident. It is well established that individual errors occur during aviation operations; however, such errors seldom propagate to the point of an accident because of the many safe guards built into the aviation system. One of the important programs developed to mitigate individual errors is CRM training and procedures. The ICAO Human Factors Training Manual (Doc 9683) states (Part 2, paragraph 2.2.9), *"CRM is a widely implemented strategy in the aviation community as a training countermeasure to human error. Traditionally, CRM has been defined as the utilization of all resources available to the crew to manage human error."*

The ICAO *Human Factors Training Manual* also states, in part *(Part 1, paragraph 1.4.25), "Crew coordination is the advantage of teamwork over a collection of highly skilled individuals. Its prominent benefits are:*

- *an increase in safety by redundancy to detect and remedy individual errors; and*

- *an increase in efficiency by the organized use of all existing resources, which improves the in-flight management."*

One of the basic elements of CRM involves checklist discipline. The general concept involves one pilot performing a check, while the other pilot confirms or monitors to ensure that the proper actions have been taken.

The ICAO Manual further states, in part *(Part 1, paragraph 1.4.26), "The basic variables determining the extent of crew coordination are the attitudes, motivation, and training of the team members. Especially under stress (physical, emotional, or managerial), there is a high risk that crew coordination will break down. The results are a decrease in communication (marginal or no exchange of information), and increase in errors (e.g. wrong decisions), and a lower probability of correcting deviations either from standard operating procedures or the desired flight path ... ".*

The Manual adds *(Part 1, paragraph 1.4.27),* in part, *"The high risks associated with a breakdown of crew coordination show the urgent need for Crew Resource Management training, ... This kind of training ensures that:*

- *the pilot has the maximum capacity for the primary task of flying the aircraft and making decisions;*

- *the workload is equally distributed among the crew members, so that excessive workload for any individual is avoided; and*

- *a coordinated cooperation - including the exchange of information, the support of fellow crew members and the monitoring of each other's performance - will be maintained under both normal and abnormal conditions."*

According to training records, the Captain and First Officer had received CRM training in June 2005 and February 2005, respectively. Records also showed that the cabin attendants had received CRM training in 2005. Nevertheless, the Board believed that the circumstances of the accident indicate a breakdown in crew coordination and inadequate CRM that did not mitigate individual errors made at several stages of the operation.

During execution of the initial preflight duties checks by the flight crew and during completion of the Preflight Checklist, the flight crew did not set and verify that the pressurization mode selector was in the AUTO position. The First Officer had the primary responsibility for the proper setting; however, the Captain had a role to oversee and monitor the correct actions by the First Officer, particularly during the execution of the Preflight Checklist. That is the essence of good CRM.

Similarly, the flight crew did not notice the incorrect position of the pressurization mode selector during completion of the After Takeoff checklist. The responsibility for the pressurization setting rested on the pilot monitoring; however, the pilot flying had a responsibility to confirm the proper setting, which would occur if proper CRM was practiced.

When the pressurization warning horn sounded, as the aircraft cabin altitude climbed above 10.000 ft, the flight crew also exhibited inadequate CRM as they were not able to determine the true reason for the warning, and they did not silence the horn. Consequently, they did not stop the climb and they did not don their oxygen masks. The Captain exercised good CRM by contacting the Operator on the radio for assistance during the climb. However, confusion during those conversations and the likelihood of the onset of hypoxia precluded satisfactory resolution of the situation. The flight crew of HCY522 did not exhibit adequate CRM to help overcome the individual errors and to detect a dangerous situation that deteriorated as the aircraft continued to climb.

It is well established that scores of past airline accidents and incidents have been associated with pilots' lack of adherence to proper checklist procedures. The record of past pressurization incidents and accidents examined by the Board reflected that a significant percentage of the occurrences involved non-adherence to checklist procedures. One of the important reasons for adherence to good CRM practices is to ensure checklist discipline and to make effective use of existing resources to improve safety of flight, by using the "team" approach to overcome inherent human errors. If one of two pilots in a cockpit displays less-than-optimal discipline in performing checklists and is not as effective in dealing with non-normal situations (as documented, at times, in the First Officer's training record), good CRM would have enabled a two-pilot crew to function as a team in order to avoid inadvertent omissions, to rectify them as soon as possible, and to effectively and swiftly manage non-normal situations.

Good CRM also involves cooperation between the flight crew and cabin crew for certain non-normal events, particularly pressurization problems. In this occurrence, it was not possible to determine what actions, if any, were taken by the cabin crew to gain the attention of the flight crew after the passenger oxygen masks deployed and the aircraft did not level off or begin a descent. For many non-normal events, the cabin crew plays an important role in the team resolution of developing problems. For example, in one of the incidents investigated and reported by the Irish AAIU, *"The continued persistence by the (Senior cabin attendant) in keeping the flight crew informed of the cabin situation was a major factor in ensuring the safe outcome of this serious incident."* In that case, proper CRM practices overcame the individual errors that precipitated the event.

Scores of aircraft accidents have occurred in the past because of inadequate CRM. Many of the previous pressurization incidents examined by the Board revealed individual errors and initial inadequate CRM that caused the initial problems; however, subsequent proper reactions by the flight crew, including good CRM, prevented the events from leading to an accident.

Cabin Altitude

Based on calculations performed by Nord-Micro on data from the NVM, the cabin altitude during cruise was 24.000 ft. According to calculations by the Boeing Company, the cabin altitude was between 20.500 ft and 28.200 ft. At these altitudes, and based on the TUC table, occupants on board not wearing oxygen masks started experiencing a gradual decline of cognitive functions around 2 - 3 minutes after the aircraft acquired FL340.

Operator

The required manpower to cover the maintenance activities in Larnaca (LCA) consisted, in accordance with the Maintenance Management Exposition (MME) of the Operator, of four licensed engineers and four mechanics, who together with the rest of the new maintenance management personnel, must have completed an initial training to become familiar with the Operator's internal policies and procedures, legal obligations and relevant legislative (JAR OPS 1 / Part 145) requirements. Continuation training on several important maintenance issues would follow at intervals not to exceed two years, or more frequently at the discretion of the Technical Manager. All engineers must meet ATC Lasham authorization requirements, as far as their license was concerned (paragraph 0.3.8, Manpower Resources and Training Policy, Issue: TR 2, Date: 17.08.2004).

Based on evidence from the Helios Airways Technical Department documents relevant to manpower planning, the front line maintenance task force group consisting of four to five licensed engineers and two to three mechanics, changed, as far as the individual persons of the first group were concerned, by more than 80 % three times within 16 months (oldest EMPLOYMENT DATE: 01/11/2003 – first END DATE: 04/03/2005). The longest stay with Helios Airways up to 14 August 2005 was 21 months and the shortest three days. Both example cases above were licensed engineers and categorized as *"Permanent"* in the column *"Employment Status"* of the document. The same column contained another category, *"Contract"*, reserved for those licensed engineers hired through employment agencies.

Between November 2003 and August 2005 (one week before the accident occurred), 13 licensed engineers (six different European nationalities) were employed by the Helios maintenance department and subsequently left Helios. Six of them were contracted, which meant that they were paid by the employment agency that placed them with the airline. The Board believed that the very high turnover rate of maintenance personnel was not conducive to establishing and maintaining a sense of continuity and teamwork among employs, and this probably worked against setting a good foundation for proactive management and resolution of any issues in the maintenance department.

The situation at the Larnaca line station as far as manpower planning was concerned had also been raised as a Non-Conformance Report (NCR) during an audit carried out by ATC Lasham on 6 April 2005. The certifying staff level (number of licensed engineers), not including the Maintenance Manager, was annotated in the NCR as insufficient to meet the requirements of Part 145. The manpower plan appeared to only be a guide and did not fully reflect the current status of manpower usage or requirements. The corrective action to the NCR from the Operator's maintenance management stated that as of 1 April 2005, the policy of the Operator's Engineering & Maintenance was to only employ licensed engineers with the relevant type ratings. Mechanics with a license, currently with the Operator, would be offered type training and be bonded to the company for a three year period. In addition to the above, the Operator had an agreement with Cyprus Airways for maintenance engineers on an "as and when required" basis to cover line maintenance peak requirements. Despite the above intention of the maintenance management to improve the situation, the responses by airline management continued to prove inadequate to provide for the necessary resources and financial support.

The licensed engineer who was responsible for the preparation of flight HCY522 in the early morning of 14 August 2005 was also hired through an employment agency (TAC Europe) and authorized by ATC Lasham in accordance with the contract between the maintenance contractor (ATC Lasham) and the airline (Helios Airways) as JAR OPS 1, Subpart M prescribed. The licensed engineer had also been employed by Helios approximately two years earlier for the duration of eight months (from 25 October 2002 to 20 June 2003). This second contract period started on 15 April 2005. There was no evidence that the engineer had participated in a refresher familiarization course as prescribed by the Helios MME, and this fact was confirmed by the engineer. In fact, he stated that he was not aware of the existence of any such courses. He was an experienced and capable engineer. According to his Curriculum Vitae, he was employed by the UK Royal Air Force for seven years and with two well known British private aviation entities for the next five years. He started to work as a freelance engineer in 1990. It was assumed that he had been part of an established and well-maintained company safety culture at least during the first 12 years of his professional career when working for major UK operators. He was the holder of a UK CAA Aircraft Maintenance License with more than 200 Aircraft Type Rating endorsements, the vast majority of which were issued on 2 September 2002 and on 5 July 2004.

The engineer departed Cyprus after having made a written statement to the Larnaca Police on 29 August 2005. After he was requested to appear before the Board to submit an oral statement, he communicated with the Helios Maintenance Manager expressing his inability to be available in Cyprus. However, he stated that he was willing to meet and assist the investigators in the UK.

The other engineer, who assisted the above discussed engineer in the pressurization leak check on 14 August 2005, had been working for Helios Airways for about six weeks on release from ATC Lasham. The same morning, he left as scheduled to return to his base in Lasham, UK. He had also been with the Operator in the past, from June to November 2003. At that time, he had been hired by the employment agency TAC Europe.

The former Technical Manager of the Operator was asked why he resigned in January 2005 after having served the company for more than four years. He answered that the reason for his decision was the mismanagement in cases such as:

- Staffing of key posts e.g. Quality Manager, Flight Operations Manager, with individuals who either did not have the required qualifications by the Operator's Policy prerequisites, or did not possess managerial competence;

- Lack of business planning;

- Incoherent corporate operations; and

- Occasional coverage of personnel requirements in all specialties of the corporate operations.

The findings related to aircraft maintenance in the audits conducted in 2004 and 2005, as discussed in 1.17.3.3 and 1.17.3.4 above, indicated efforts by

ATC Lasham to obtain the necessary technical documentation and completed maintenance records from Helios Airways covering the maintenance activities in Larnaca. The Board noted that the Technical Administrator of ATC Lasham had repeatedly requested the completed maintenance documentation required by the contract between ATC Lasham and Helios Airways, but he appeared unsuccessful. The Board noted that he appeared to have resigned to the fact that Helios Airways may not provide the required documentation. This constitutes an unusual perspective of a responsible individual serving a technical company which has been established in the international market for more than 5 decades. According to the Maintenance Agreement, per Part 145, both parties are obliged to take direct measures for the eradication of anomalies. Non-conformance of one party to such a serious case, automatically gives the other party the right to interrupt the collaboration in order to prevent unpredictable situations. The shortcomings with the required documentation on aircraft maintenance from Larnaca were an indication of inadequate coordination and understanding of the contractual obligations between Helios Airways and ATC Lasham.

The Board also noted that the audits in 2004 of the maintenance activities conducted by a UK CAA inspector resulted in repeat findings from earlier audits, level 2 findings and also in more serious level one findings. Subsequently, all the outstanding findings were confirmed closed as a result of a meeting between the inspector and the Technical Managers of Helios Airways and ATC Lasham, however, some concern remained with the Board as to the corrective actions required and actions taken. The Board believes that a level finding 1 must be reported as level 1. There are no other alternatives. The philosophy of level 1 findings provides the instrument to interrupt the sequence of events that may lead to an accident. Regarding the maintenance of the accident aircraft, the Board noted that the readout of the NVM chip in the cabin pressure controller showed that the NVM chip had recorded messages which indicated a continuous leakage situation with the aircraft. The recorded fault messages were *"030 INFLOW/LEAKAGE"* and had persisted for at least the last 74 flight legs. As discussed earlier, these fault messages were triggered by low flow through the OFV due to low inflow or high leakage rates in the aircraft fuselage. The Board also noted the 16 December 2004 incident when the accident aircraft experienced a rapid decompression at cruising level FL350.

The AAIIB of Cyprus had documented the incident but was not able to reach a conclusive decision as to the causes of the incident. Furthermore, the Board noted continuing problems in the accident aircraft with the equipment cooling system as evidenced by nine write-ups in the Aircraft Technical Log from 9 June to 13 August 2005. Each of the write-ups was actioned by the Helios maintenance, but it was of concern to the Board that the problems with the equipment cooling system appeared to continue despite the maintenance actions. And the problems with the equipment cooling system occurred again on the accident flight. The Board determined that the leakages had most likely persisted over a longer time period as had the problems with the equipment cooling system. The Board believed that despite the maintenance actions taken, the underlying causes for these technical discrepancies appeared to have remained unsolved. The Board considered these unsolved technical discrepancies in the aircraft to be a further indication of inadequate organizational issues within Helios Airways and its maintenance organization. Although the Board found that the technical discrepancies in the aircraft may not in isolation and by themselves have been causal factors in the accident, it was evident to the Board that at least the problems with the equipment cooling system had been the subject of the preoccupation of the flight crew in a critical phase of the flight.

Crew scheduling

According to the records made available to the Board, the crew duty times were within limits and followed the prescribed standards. However, in view of these records and a number of statements made to the Board, it had reservations on this subject, given that the records submitted required extensive examination to validate flight and duty times for the flight crew. The Board noted that inspectors/auditors in previous audits had annotated comments that the Captain's Deviation Reports (CDRs) showed flight and duty times that exceeded the approved limits and were not recorded or reported to the DCA. The Board also noted statements that the scheduling of flights was based on unrealistic flight times for some routes in order to ensure flight planned adherence to flight time limitations which subsequently were exceeded.

Crew Training

Based on the training records reviewed, the flight crew training was approved by the Cyprus DCA and carried out in accordance with the Helios Flight Training Manual. According to the manual, the simulator training syllabus included rapid decompression situations, but not gradual decompression (slow loss of pressurization) situations. Consequently, the flight crews were likely not sensitized to monitoring and detecting a more insidious, gradual loss of pressurization situation. The Board noted that the Norwegian AIB had promulgated a safety recommendation (no. 27/2002) to the airline involved in a loss of pressurization incident in Norway in 2001 to the effect that gradual loss of pressurization situations be included in simulator training.

The Board reviewed the Helios Airways Training Manuals and identified a specific requirement for training of both flight and cabin crews on the phenomena associated with hypoxia. However, based on witness statements, the Board was led to believe that this requirement was not fulfilled in practise but remained a requirement "on paper." The Board noted that this situation was not unique to Helios Airways, because the lack of hypoxia training to sensitize flight crews to detecting an insidious gradual decompression or non-pressurization of the aircraft during climb, was a common situation in the airline industry.

Furthermore, the Board noted that hypoxia training tools had recently been developed and were readily available and produced in several countries. These hypoxia training tools can demonstrate many of the effects of hypoxia by gradually reducing the amount of oxygen a pilot receives while wearing a training tool mask and performs tasks. These tools are used by the US Navy to provide flight crews with realistic hypoxia training without the inherent difficulties and costs of providing altitude chamber training. A description of the new technology and the manufacturers can be found by searching the Internet for "hypoxia training aids".

Interviews with a number of cabin crew members (including Cabin Chiefs) revealed a number of deficiencies. In particular, cabin crews appeared confused and responded differently to questions that concerned the number and type of oxygen masks on the B737 flight deck, the availability and exact procedure of means available to open the cockpit door, and whether passenger oxygen masks provided breathable oxygen at high altitude. Furthermore, deficiencies were also identified in the

Operator's procedures that prescribed actions to be taken in the event that, after passenger oxygen mask activation, the aircraft did not begin to descend or at least to level-off. However, it was also determined that other airlines in Cyprus and in Greece did not have such procedures documented in their manuals.

As far as access to the flight deck was concerned, the Board considered the procedures available to open the flight deck door. According to cabin crew statements, upon upgrading to Cabin Chiefs, they became aware of the appropriate procedures. Only one of the interviewed Cabin Chiefs had actually used the procedure but could not recall whether the door was electrically powered or not. Guidance on the procedure was contained in the Helios Flight Safety Manual (Chapter 5, page 5-34). According to the manual, the procedure was useable only when the door lock mechanism was inactive (either by choice or due to an electrical failure). Even then, the pilots in the cockpit could use the deadbolt and position it such that access to the cockpit was not possible.

The Board considered the flight deck emergency access training deficiencies and inconsistencies, and on the other hand, it took into account the apparently random and risky practice of distributing copies of the emergency procedure. Although some aspects of the procedures at Helios could be considered unsafe, the Board determined that these issues were probably not implicated in the accident.

Operator Management

The Board was concerned that the management structure at Helios Airways at the time of the accident was incomplete, notably the position of the Manager Training Standards. Helios Airways stated that a Manager Training Standards had been appointed in March 2005, but he had resigned on 25 July 2005. In these circumstances, the Flight Operations Manager had assumed the responsibilities of the Manager Training Standards, pending a re-organization of the Operations Division and the arrival of the new Chief Operating Officer at the beginning of August 2005. Furthermore, the Chief Pilot who was a qualified TRE, was in a position to deputize for the Training Manager Standards. The qualifications of some of the interviewed managers did not correspond to the qualifications listed in their job descriptions. The statement by the former Technical

Manager of the Operator confirmed the same. The Board noted that the deficiencies in management, particularly the position of Manager Training Standards, may have been related to the failure of the Operator to recognize and take appropriate corrective actions to remedy the chronic checklist and SOP omissions exhibited by the First Officer and documented in his training records.

Despite the management's assurance that the work climate at the company was healthy and the employees satisfied, a number of interviewed employees voiced complaints and concerns towards the Operator's management (other employees denied any problems with approaching the management). The Accountable Manager was characterized as unapproachable, with little regard or concern for safety or for the well-being of the company employees, and whose only interest was the profitability of the Operator.

The Board acquired the sense that the overall philosophy and style of management at Helios Airways was not conducive to efficient and safe operations. This impression was corroborated by the UK inspector's comments in July of 2004 expressing concern about the potential that flight safety was being compromised due to *"the lack of operational management control"* and the hesitancy with which some improvements were made, were noted by another inspector a year later.

Operator Staffing

The Board considered potential implications of the multi-national staff composition at Helios Airways and how they might have affected the safety of flight operations. Multi-national teams often led to a weak work climate because people of different cultural groups operated based on a set of values and perceptions unique to their common historical/social/geographical background. These types of differences might lead to communication and collaboration problems. A characteristic example was that of the picture many employees had of the accident Captain. His East German heritage meant that he was likely a man of few words who was not very comfortable around people, and thus less expressive towards his colleagues. He was also therefore likely to have been direct in his challenges and exchanges with crew members, and less flexible with deviations from procedural standards. People of a Mediterranean temperament were different in some of these aspects. The Board therefore believed that some of the Captain's Cypriot colleagues likely perceived him as authoritative, dominating, and unfriendly. In fact, the perceived discontent and complaints about the Captain's abrupt manner and his interference with First Officer duties came from Cypriot pilots – none of the non-Cypriot pilots interviewed expressed any difficulties in their collaboration with the accident Captain. The Board understood it to be a question of perception of the attitude of the Captain, and how he was perceived when addressing his colleagues.

A relevant comment of interest came from a former chief cabin attendant. According to her statement, the Captain of the accident flight was very formal and enforced company rules. She gave an example describing that the Captain required the crew to disembark the aircraft (as required by company rules) during the scheduled 6-hour stays in Tel Aviv and to make use of the airport lounge which was dirty and noisy. Contrary to Operator procedures, other Captains had usually permitted the crew to stay on board the aircraft. This type of action though completely justified since it was required by Operator procedures and policy, may have been regarded as "unfriendly" or "authoritative" by some.

Another area of concern that arose from the composition of Helios Airways staff stemmed from the large percentage (33%) of staffing with seasonal (part-time) employees. Naturally, this was expected for companies whose operations mainly catered to the tourist industry and were, by definition, seasonal. The short-term hiring of pilots and engineers when the operational tempo and demands were significantly higher in the spring and summer allowed the airline to maintain a skeleton staff to cover the less-loaded winter months. Insofar it affected work climate, however, frequent changes in staff composition could be detrimental to the development of professional and personal ties, and did not promote the required level of comfort among employees, and among employees and management, particularly with respect to the submission and discussion of incidents and problems. Employees lacked a sense of continuity, both for their own job as well as that of their colleagues, and cockpit and cabin crew did not have the opportunity to develop operational experience together in various routine and non-routine situations. Employees, finally, did not develop a feeling of ownership and responsibility towards operations and the Operator. A pilot or engineer who was on a short-term contract had less chances of feeling an important, functional part of the operation. It meant that they had limited responsibility and that it was perhaps not worth reporting or addressing potential deficiencies. It was notable that this type of work climate deteriorated further when the Operator, due to the seasonal nature of its operations, chose to lend its employees to other companies with opposite seasonal needs. Such a move could make its employees feel more dispensable and less essential to the operation.

Operator Safety Culture

Provisions existed in manuals for an accident prevention/safety management program at Helios Airways. However, it was not at all clear whether the Operator adhered to the standards set forth in the relevant publications. Furthermore, these standards seemed to promote a reactive approach rather than emphasizing the benefits of a more effective, proactive stance to safety management. More important, the standards did not clearly and definitively outline the role and responsibility of management (a key element in any safety management program) in ensuring and maintaining safe operations of the company. The Board found reasons for further concern in the statement by the Chief Operating Officer of Helios Airways. The Board recognized that he had joined Helios Airways just two weeks prior to the accident and that he may not yet have had a detailed, in-depth view of all aspects of the Helios Airways flight operations. On the other hand, the Board also recognized that the perspective of an experienced "newcomer" to the Operator can be objective and illuminating, especially in matters, such as company safety culture, that should permeate all aspects of the operations and should be transparent to a newcomer. By referring to tight schedules both for employees and aircraft utilization, the Chief Operating Officer appeared to suggest that both resources were utilized to the limits. The Board noted that tight scheduling, work under time pressure and considerable amounts of over-time work were not conducive to maintaining a safe work environment. These conditions were likely a fertile basis for human factor errors in flight operations and aircraft maintenance.

The Board's review of the available Operator audit reports showed many findings to repeatedly concern inspectors. Management pilots appeared to be insufficiently involved in their managerial duties. Training and duty records were found to be incomplete. Manuals were found to be in part deficient; they did not always adhere to regulations, and on some issues they were out of date. In the two months before the initiation of the first flight operations with the then newly-acquired B737-300 (the accident aircraft), the airline appeared to be effectively scrambling to piece together manuals and paperwork. This suggested that an underlying pressure was prevalent to proceed with little regard for the required formalities (which often equaled an assurance for safety). According to the UK CAA, before the AOC variation for the B737-300 could be recommended, the airline was required to have appropriately qualified personnel in place with the Terms of Reference defining areas of responsibility.

The Board's review of the accident flight crew's training records revealed issues noted in the First Officer's record but no evidence of any type of a follow-up, for example from the Operator's Chief Pilot or Training Manager. The Board considered such issues (insufficient checklist discipline, tendency to not perform effectively in non-normal situations) to be quite disconcerting and believed that they should have warranted some type of action on behalf of the Operator. The periodic vacancy of the position of Training Manager, coupled with the fact that various auditors in the recent past had noted the failure of management pilots to spend sufficient time at the office, led the Board to note that the Operator lacked the mechanism and means to sufficiently and correctly monitor its pilots and to take decisive and corrective action when and as necessary.

Lastly, the Board also reviewed the actions of the Ground Engineer team that conducted maintenance on the aircraft prior to its departure, so as to form an opinion about the operation of the maintenance department at Helios Airways as a whole. The inexplicable inconsistencies in the actions that were or were not performed, the actions recorded, and the actions described as having been performed by Ground Engineer No. 1 on the morning of 14 August 2005 were considered by the Board to confirm the idea that the Operator was not effectively promoting and maintaining basic elements of safety in its culture

Operator Quality Assurance

The Board considered the issues identified above to be critical for any system to operate safely, as well as efficiently. Such issues, furthermore, might have been addressed internally, before the audits, or at least promptly after each audit, had the airline had an effective, functional Quality System. The Board echoed the DCA's sentiment that such a system was lacking at Helios Airways. The delay in appointing a Quality Manager, the failure of the airline to furnish a quality audit plan, audit reports, or other related documents, and the non-availability of any documents to suggest the internal management evaluation was taking place as required, were all proof for that.

It may be worthwhile to note that the three appointed quality auditors were the Technical Pilot, the QM himself, and the Technical Manager. According to the Quality Manual, *"Auditors should not have any day-to-day involvement in the area of the operation and/or maintenance activity which is to be audited."*

That quality assurance was not effective in specific cases at Helios Airways was evident to the Board who was often unable to establish the most recent revisions of certain manuals. The official approval of the Cyprus DCA was not evident on the manuals submitted to the Board. The Operations Manual, Part A which was the Flight Operations Manager's responsibility, appeared to be in need of revisions when the airline was audited in 2004. During the investigation, further, detailed examination of the Operator's manuals failed to yield sufficient evidence that they were being properly and/or regularly updated. For example, the QRH containing the Normal Checklists contained the x, 2005 version of the After Takeoff checklist, although the manufacturer had issued a revision in X, 2005 that should have become incorporated in the manuals of all companies flying the particular aircraft. Specifically, the older version of the After Takeoff checklist in the Operator's QRH displayed the following as item x of x: *"AIR COND & PRESS ..... SET"*. The revised version had changed this to two separate items *"ENGINE BLEEDS ON"* and *"PACKS AUTO"*. When asked about this during the investigation, the Flight Operations Manager stated that although the Operator received updates through the aircraft's previous owner (who continued to have a contract with the manufacturer to receive such updates), the Operator exercised their own screening of updates and selectively applied only those they considered *"important"*. The Board was particularly concerned about the standards based on which an update (i.e., a change to a checklist item) was deemed not important enough to take into account. Though not a causal factor in this accident, the failure to incorporate official changes in manuals was considered by the Board to be a display of insufficient quality assurance, with a potentially significant impact on safety.

Air Traffic Control

Nicosia ACC

During the climb phase of the accident aircraft, and after losing radio contact with it, the Helios Operations Centre Dispatcher asked Nicosia ACC to contact the accident flight. After six calls to which it received no response, the controller acted in accordance with the provisioned procedure to check whether the flight had two-way RCF.

According to the controller's statement to the Board, at this point Nicosia ACC realized that there was a two-way RCF situation. The Nicosia controller informed Athinai ACC without using the formal procedure in ICAO Doc 4444, paragraph 15.3.7 and paragraph 11.4.1.3.1, i.e., by telephone and then by sending a formal RCF message through AFTN. Instead, the controller communicated with Athinai ACC two times via telephone, stating *"Helios at EVENO does not respond. If he calls you, let us know"* and *"Did Helios call you?"* The Board considers this transmission of information to Athinai ACC regarding the status of the accident flight to be sub-standard. In addition, the Board believes that Nicosia ACC never fully realized that there was a two-way RCF situation.

Athinai ACC

One minute before the flight entered the Athinai FIR, the color of its target was automatically changed by the radar system from green to salmon color. The Athinai Radar Controller "clicked" on the target and changed its color from salmon to blue denoting that the flight had been acknowledged and "accepted." Simultaneously, the system automatically coupled the target, when it was at the FIR boundaries with its flight data that had been received earlier by written ESTIMATE message (i.e., estimated time of arrival at EVENO, flight level, and transponder code).

Given that the flight, however, was still flying within the Nicosia FIR and was still on that frequency, at the moment he intervened on the flight's target and changed its color from salmon to blue, the Athinai Radar Controller did not call it, but waited for it to arrive to EVENO above which the flight was supposed to give a report about its entrance as per the provisions in the AIP Greece.

The flight finally entered into the Athinai FIR without contacting the Athinai ACC, because of the condition in which its flight crew was in.

Given that:

- the previous action of acknowledgement and acceptance of the flight, and the fact that the Controller could not call the aircraft at that moment because it was still flying within the Nicosia FIR and was therefore still on the Nicosia ACC frequency;

- that the flight crew was not able to communicate with Athinai ACC upon entering the Athinai FIR, per procedure, due to its condition;

- none of the flight data changed on the radar screen nor did any indication or message appear to show that there was any problem (e.g. transmission in one of the three emergency codes, 7500, 7600, or 7700);

- the Radar Controller had not been informed by anyone that there was a loss of communication with the a/f, because according to his statement, the Planner Controller of Athinai ACC he did not inform him about the two telephone calls from the Nicosia ACC, that there was no contact with HCY 522. Of course, in his statement the Planner mentions that *"I tell (the Radar Controller) that we don't have any contact with the flight, without receiving confirmation that he heard me."*). However, the Radar Controller said in his statement that *"I was never given that information".*

After the above, the Radar Controller, under the impression that he had already called the flight when he had intervened on its target changing its color and its label, upon its entrance into the Athinai FIR and the fact that at the same time there were other flights calling him and with which he was communicating, considered the flight to be progressing normally and did not call it. According to ICAO Doc 4444, paragraph 9.2.2.1, the Controller should have called the flight when, upon entering the Athinai FIR did not call, as it was required to do per AIP Greece, RAC 1-1-3, paragraph 2.1.6.2. Paragraph 9.2.2.1 of Doc 4444 mentions that "... *when no report from an aircraft has been received within a reasonable period of time (which may be a specified interval prescribed on the basis of regional air navigation agreements) after a scheduled or expected reporting time, the ATS unit shall, within the stipulated period of 30 min, endeavour to obtain such report in order to be in a position to apply the provisions relevant to the "uncertainty phase" should circumstances warrant such application*" and paragraph 2.1.62 of RAC 1-1-3 of AIP Greece mentions that "*a position report, shall be transmitted when entering Athinai FIR boundaries from IFR aircraft*". Therefore, the ICAO regulations do not explicitly specify the time interval after which an Air Traffic Service will have to call a flight that omitted to report its position, but mentions that this must be done in a "*reasonable period of time*" that may be defined by regional aeronautical agreements. The Athinai ATS had not specified this time interval by any regional agreement.

The Radar Controller may not have called the flight, but even if he had after its entrance into the Athinai FIR around 09:37 h, the flight, because of the condition in which its flight crew was in would not have responded. In that case, the Radar Controller, upon realizing the RCF would not have taken any direct action after 30 min (10:07 h), that is to declare an "*Uncertainty Phase*" because at that moment in time (09:37 h) the flight appeared to progress normally and did not show any change in its information that had been received earlier by Nicosia ACC by written message and verbally by telephone. I.e.:

- the flight was at FL 340 as instructed by Nicosia ACC

- its transponder code that accompanied it was also the same without any change

- its estimated time of entry at reporting point EVENO coincided with the real time that the flight flew over it

- the flight's path was normal without any deviations from the track of the airway

- the target on the radar screen appeared without any changes, and with no change in the transponder code that it had been assigned by Nicosia ACC to any one of the 3 emergency codes (7500, 7600, 7700).

-

According to Annex 11, paragraph 5.2.1 which mentions that "... *air traffic services units shall ... notify rescue coordination centres immediately an aircraft is considered to be in a state of emergency in accordance with the following:*

1.   *Uncertainty phase when:*

2.   *no communication has been received from an aircraft within a period of thirty minutes after the time a communication should have been received, or from the time an unsuccessful attempt to establish communication with such aircraft was first made, whichever is the earlier ...except when no doubt exists as to the safety of the aircraft and its occupants.*"

and given that the Radar Controller, according to the above, had no doubt about the safety of the aircraft and those on board, did not take any action to declare an "Uncertainty phase."

The only case he could have suspected at that moment in time was RCF with the aircraft, something which has happened in the past and for this reason ICAO has instituted procedures that the ATS must apply and the flight crews must follow. According to these procedures, the flight must remain at the last flight level it has been flying at (FL 340) and must head to the appropriate waypoint (KEA/VOR) that services the destination airport. Upon reaching that waypoint (HCY 522 reached it at 10:2 1 h), it must start a normal descent following the approved standard approach procedure published for the specific waypoint and which the flight crew must know, and then must continue for landing. Until then at 10:2 1 h the Radar Controller would not take any further action because the flight remained normally at FL340 and was headed to the specific KEA/VOR waypoint from which it should have started a normal descent.

In both cases, therefore, the Radar Controller, whether he called the flight at 09:37 h when the flight flew over reporting point EVENO and he had established a RCF, or when he called it at 10:12 h shortly before it flew over RIPLI to give the flight clearance for the descent where he established a RCF, he would expect the flight crew to follow the procedures prescribed by ICAO Doc 4444, paragraph 15.3.3 (b) and he would not take any other action except to alert the other Air Traffic Services, that is Approach Control and Athens International Airport Tower, which he did.

The flight remained normally at FL340 and the Radar Controller monitored its progress, if it was being conducted per regulation. Therefore, there was no negligence or delay in alerting the responsible parties until that moment in time. When at 10:2 1 h, however, the flight flew over the KEA/VOR and although the flight crew should have initiated the standard approach procedure, the flight remained at FL 340 continuing to not respond to the repeated calls by Athinai ACC, the Radar Controller revisited the case of an RCF with the flight, notified the Athinai ACC – whom he had earlier informed about the loss of communication with the a/f – that something different was now happening. The NOC was notified, as was the JRCC and any other involved party. Following that, an Alert phase was declared when the flight flew over the Athens International Airport, executed a missed approach towards the KEA VOR while remaining at FL 340 and entered the holding pattern. Finally, a Distress phase was declared when the F- 16 pilot reported that there was no one in the cockpit.

Department of Civil Aviation in the Republic of Cyprus

At the time of the accident and as emerged from employee oral statements and written reports that were made available to the Board, the Safety Regulation Unit (SRU) was diachronically not organized and staffed to effectively accomplish its regulatory and safety oversight duties. The main problems that characterized the Unit and each of its three Sections (Operations, Airworthiness, and Licensing) already back in 1999 (the time of the first available audit report) appeared to still persist to this day, as evidenced by the various evaluation reports reviewed. The number of employed personnel was insufficient in relation to the actual workload. The mission and strategy of each Section, including its processes and standard operating procedures, appeared not to be officially laid out in writing. Selection and training criteria and resources, as well as detailed job descriptions were not available. By extension, the qualifications, training, and hands-on expertise of most employees were probably inadequate. Vital positions (e.g. Head of the Operations Section) remained vacant. Some key functions (e.g. issuance and validation of air transport pilot licenses; issuance and record-keeping of medical certificates) were not performed. Other key functions (e.g. inspections) were possibly not accomplished per schedule because qualified personnel were not readily available and external resources had to be relied on.

This diachronic absence of leadership and oversight both across and within the three Sections presented a major obstacle that hindered the effective work of any one of the Sections. The resulting work climate within the SRU was not conducive to good performance even by qualified personnel; this became apparent in the nature of the oral statements given by the employees that included charges and complaints, as well as direct accusations and finger-pointing. Given the situation within the SRU, it was probably difficult for the Unit to instill a level of esteem from the aviation industry, and specifically in the areas and activities that it was tasked to regulate and oversee.

To accomplish its safety oversight duties, the Unit relied heavily on the UK CAA to furnish (based on a contractual agreement) inspectors to carry out the ICAO and EU required inspections. Based on the contractual agreements between the Cyprus DCA and the UK CAA, the role of the latter was undoubtedly intended to be advisory in nature.

In reality, however, the DCA appeared to have been fostering and maintaining a relationship of complete dependence on the UK CAA, and, in most cases, appeared to be simply accepting its services without questioning them and without making an effort to assume ownership and thus build on them. The Board was particularly concerned to find that almost all of the Operator's audit reports until about the time of the accident were signed by the UK CAA inspectors without any comments and/or a signature by an employee of the Cyprus DCA. Where the situation at the audited Operator seemed to repeatedly yield deficiencies and issues that required often urgent attention, the Board found no evidence that the DCA would actually "step in" and take action to ensure that the Operator complied and took corrective actions and, consequently, was safe and legal to continue its flight operations. As mentioned in the evaluation by a private firm in 2005 *"The UK CAA representatives acknowledged that their current role in Cyprus is as advisors. However, this remains unclear since the existing contracts indicate, and records confirmed, that the UK CAA inspectors exercised a more direct, "hands-on" approach."*

The relationship of dependence was also evident from the evaluation by the private firm which found that the DCA had not taken ownership of documents prepared by the UK CAA and which described the internal operations of the Flight Operations and the Airworthiness Sections of the SRU. The SRU appeared to have adopted the manuals without completing missing sections and/or tailoring them to their needs, or trained inspection personnel to use them.

In trying to explain the reasons behind the slow progress in strengthening the DCA capabilities, the Board considered the role of Governmental support and how that may have been affecting the DCA's ability to evolve and better embrace its safety oversight responsibilities. The Board noted that the 2002 ICAO audit clearly attributed at least part of the situation to the fact that the DCA operated as a functional department of the Ministry of Communications and Works. The 2005 European Commission evaluation directly faulted the absence of the necessary *"... political commitment* (of the Cypriot Government) *to supply this Department* (DCA) *with the resources to carry out fully its safety oversight function and to reorganize the chain of command in order to give safety the high priority it deserves inside the organization."*

What became apparent from the Board's consideration of the situation at the Cyprus DCA, and what was evident from the review of the audits/evaluations of DCA, was that the DCA, and the SRU in particular, lacked the required expertise to move forward, become independent, and fulfill the international obligations of Cyprus as contained in the Chicago Convention and its Annexes. Despite numerous action plans since 1999 to ensure the availability of properly trained and qualified inspectors, there were no tangible indications of progress, except for the UK inspector's written statement that *"..... audits have been increasingly performed in conjunction with the DCA 's technical staff to provide on the job training, thereby preparing them to perform these inspections in the future without the external support of the UK CAA."*

According to the UK CAA, any delays which occurred in the provision of on the job training (OJT) arose from reasons outside the control of the UK CAA. The UK CAA cited instances in which Cypriot flight operations inspectors had been assigned to an UK CAA Overseas Flight Operations Regulation Course in 2003 and scheduled for OJT on several occasions in 2004, but the Cypriot inspectors had failed to attend or attendance had been sporadic. This lack of success to provide on the job training to Cypriot inspectors was for reasons beyond the control of the UK CAA, thus, in this particular case, the Cyprus DCA did not receive this specific service from the UK CAA The Board noted a letter by a UK CAA inspector signing *"for Director of Civil Aviation"* (March 2005) without an indication that such a delegation of authority was, in fact, provisioned for. On the contrary, the Board noted that in a November 2004 statement, the Minister of Transports and Works of Cyprus had stressed that *"the role of the UK CAA consultants must be clearly set on an advisory basis and documents referring to various subjects must be signed by the Civil Aviation Department* (of Cyprus) *in accordance to the existing Legislations and Regulations."*

The Cypriot State, since November 2003 had proceeded in a number of actions towards reinforcing the SRU by hiring two qualified inspectors, completing the composition of the necessary JAA manuals, renewing contract of the UK CAA SRU consultants, and in June 2005 signing an agreement with EUROCONTROL for consulting services in view of the upcoming ICAO ATC audit in May 2006.

ICAO, EASA, JAA Audits

A review of the audits and follow up audits of Cyprus DCA performed by ICAO, EASA and JAA, disclosed several important findings, which should have been actioned in the shortest possible time. No records were obtained that would have documented any remedial action considered, initiated or completed. It was of concern to the Board that there was no evidence of actions and enforcement by the international regulatory agencies to require timely implementation of an acceptable action plan, although they had clearly established that Cyprus' international obligations were not being met.

Actions Taken as a Result of Previous Incidents/Accidents involving Pressurization Problems

Following each of the pressurization incidents, the Ireland AAIU issued safety recommendations that dealt with pilot procedures, checklist requirements, training, equipment design, and CRM principles. These included:

- adding a new paragraph on hypoxia in the Operations Manual to describe the insidious threat it poses to aircrew (SR20 addressed to the operator – no response received to date);

- amending the QRH to include the appropriate procedure to accomplish at the onset of a Cabin Altitude Warning horn (SR22 addressed to the operator – no response received to date);

- installing, in addition to the existing Cabin Altitude Warning horn, a visual alert warning of excessive cabin altitude (SR23 addressed to Boeing who responded on 20 June 2003: *"The cabin altitude warning is an interrupted horn that sounds when the cabin altitude exceeds 10 000 feet. Boeing does not provide a "CABIN ALT" or "CABIN ALTITUDE" warning light associated with the cabin altitude warning horn on 737 airplanes. There are no provisions for such a light nor are there any plans to offer such a light.*

- improving the After Takeoff checklist to include AIR CONDITIONING & PRESSURIZATION as a challenge and response item (SR05 addressed to the operator – accepted);

- enhancing SOPs to include a specific check of the pressurization system when passing 10 000 ft (SR06 addressed to the operator – accepted);

- improving training to emphasize the need for monitoring and recognition of insidious pressurization failures, and to remind flight crews of the need to don oxygen masks when such failures were suspected (SR33 addressed to the operator – accepted; SR34 addressed to the operator – accepted; SR39 addressed to the operator – accepted; SR40 addressed to the operator – who responded that training already contained provisions to practice such emergency procedures); and

- enhancing training to emphasize the need for communication between cabin crew and the flight deck (SR41).

Following a 15 February 2001 incident in Norway, the affected operator addressed the safety recommendation (by the Norwegian AIB, no. 28/2002) to Boeing drawing Boeing's attention to the B737-700 incident with both pack switches in OFF position, the dual use of the warning horn, and the use of a warning light for low cabin pressure in MD-80 aircraft, which triggers the MASTER WARNING light. On 9 December 2002, Boeing (Customer Service) responded as follows: *"Boeing has been receiving an increased number of reports of flight crews not configuring the pressurization panel correctly. As a result, we recently changed the Non-Normal Checklist AUTO FAIL/UNSCHEDULED PRESSURIZATION CHANGE to direct the crew to ensure the Engine Bleed Air Switches are ON and the Pack Switches are ON in the event the cabin altitude warning horn sounds.*

The high cabin altitude warning horn itself has not changed since the 737 first delivered in 1967. However, in response to your comments and similar comments from a number of airlines, we are considering development of a limited crew messaging system for the 737NG. This system would have the ability to display certain messages on the upper center display unit. One of the candidate messages would inform the crew that the pressurization system is not configured correctly (for example, PACKS OFF). This proposed system would be optional, and is at least two to three years in the future."

Another of the AIB-N recommendations (no. 27/2002) called for an evaluation to include gradual loss of cabin pressurization in the B737-700 simulator training, which the operator implemented.

Safety Recommendations following Accidents

As a result of the investigation into the 1999 accident of the LearJet35 aircraft, the NTSB addressed 11 safety recommendations (A-00-109 through A-00-119) to the Regulator (US Federal Aviation Administration). Most of these applied to scheduled, passenger-carrying operators of pressurized aircraft and included recommendations to:

- Incorporate revised guidance and information on the effects of hypoxia on human performance in training and manuals (A-00-109);

- Undertake a multidisciplinary study to determine the benefits of mandatory hypoxia awareness training (e.g. altitude chamber training) for civilian pilots and, if necessary, establish appropriate training requirements (A-00-110);

- Require operators to provide guidance on the importance of a thorough preflight briefing of the oxygen system (A-00-111); and

- Require that the donning of oxygen masks by the flight crews become a first and immediate action item in a clear and explicit emergency procedure associated with the onset of the cabin altitude warning and that relevant guidance be issued directly to pilots within 6 months (A-00-113 and A-00-114).

Most of these items were now considered closed, after the FAA published an Advisory Circular (61-107A, *Operations of aircraft at altitudes above 25 000 ft MSL and/or MACH numbers greater than .75*) in 2003, although the NTSB still considered some of the published information about hypoxia effects incomplete, and had yet to receive a report on the findings of the hypoxia awareness training study.

The investigation into the 2000 accident of the Beech King Air in Australia revealed inadequacies in the design of the warning systems in this and other types of aircraft. The Australian Transport Safety Board issued one recommendation (No. R20000288) to the Regulator (Australian Civil Aviation Safety Authority) to mandate the fitment of an audible warning to accompany the three existing visual indications designed to alert the crew to a cabin pressurization problem (i.e. cabin pressure altitude of above

10.000 ft). In response, CASA embarked on a "National campaign to cut aircraft pressure problems" and issued a Notice of Proposed Rule Making which it opened for comments. The aviation community's response opposed CASA's intention, and CASA decided not to mandate the aural warning – it instead issued a notice to operators emphasizing the benefits of such warnings and strongly recommending its installation. The ATSB considered this an open item it continued to monitor.

Alert Bulletins

On 14 December 2004, the NASA ASRS Office issued an Alert Bulletin entitled *"Boeing 737-300 Cabin Altitude Warning Horn"*, addressed to the Boeing Company and copied for information to numerous FAA offices and international aviation industry organizations. The Bulletin reviewed three incident reports. In one of the reports, the crew reported having experienced confusion in discerning between a cabin altitude warning and a takeoff configuration warning horn when it sounded while in flight. The pilot wrote *"A safety issue I would like to raise awareness about based on my experience is the lack of wisdom in having the TKOF warning horn double as the ALT warning horn. Because the cabin was losing pressure slowly, we did not feel any pressure changes in our ears that would have normally served to alert us to a pressurization prob. If the FO had not happened to remember that the horn also serves as a cabin ALT warning horn, we may have continued trying to troubleshoot the air/ground prob, until passing out from lack of oxygen."*

On 16 December 2005, the NASA ASRS Office issued a second Alert Bulletin entitled, "B737 Series Pressurization Incidents", as a follow up to the earlier Alert Bulletin of 14 December 2004. Two of the three *"reports cited in this bulletin involved initial confusion in correlating the takeoff warning horn with a cabin altitude warning"*, as follows:

- "A B737-300 first officer reported that a rapid descent was made after a cabin pressurization loss at FL330. The reporter stated that he momentarily thought the cabin altitude warning was the takeoff configuration warning horn and suggested that an indication such as a light or discrete horn would assist the crew in detection of gap loss of cabin pressure events."; and

- "A B73 7-800 captain reported that, following a bleeds-off takeoff, the crew failed to turn the bleeds on during climb and received a cabin altitude warning at FL210. The reporter suggested that there should be a verbal warning specifying "cabin altitude" or a master caution annunciation."

The ASRS office had not received any official response to this Alert Bulletin to date. The Director explained that it was not typical for the NASA-ASRS office to receive direct responses but that the industry typically used such Alerts to proceed to corrective actions.

Actions Taken by Boeing

Boeing reported that, prior to the Helios Airways accident in August 2005, it had reviewed a number of the pressurization incidents and had taken, or was in the process of taking, several actions for the B737 fleet. These actions were intended to reduce the likelihood of occurrence of particular items involved in the incidents. The actions taken by Boeing included:

- In 1999, Boeing noted increased reports on pressurization incidents on the B737 fleet. Boeing studied these reports and developed a number of improvements affecting various models of the B737. On 12 January 2000, Boeing released message M-7200-00-00139 which documented these changes and encouraged operators to incorporate them in their fleets;

- Revision to the B737 AUTO FAIL and UNSCHEDULED PRESSURIZATION CHANGE non-normal checklists to combine these procedures and to explicitly select engine bleed air switches ON and pack switches ON as the first two steps in the new procedure. Revised checklists were released in April 2002;

- Release of B737 Flight Operations Flight Crew Information Bulletin, 737 IB 2003-01, to "*inform flight crews to don oxygen masks as a first and immediate step when the cabin altitude warning horn sounds.*" This bulletin was released in February 2003 in response to FAA AD-2003-03-15 and contained an advance notice of the following revised Non-Normal Checklist: CABIN ALTITUDE WARNING HORN/RAPID DEPRESSURIZATION. The Background Information section of this bulletin included the following statement: *"There is some concern that flight crews may not recognize the horn as an alert of excessive cabin altitude. The FAA feels that revising the Airplane Flight Manual (AFM) and procedures to include reference to the cabin altitude warning horn will increase awareness and prevent incapacitation of the flight crew due to lack of oxygen, which could result in loss of control of the airplane."*

- Revision of the B737 RAPID DEPRESSURIZATION non-normal checklist to change the title to CABIN ALTITUDE WARNING OR RAPID DEPRESSURIZATION and to include as a first step in this non-normal checklist an explicit instruction to don oxygen masks. Revised checklists were released in June 2003 for the B737-300.

- Revision of the B737 After Takeoff checklist to explicitly confirm the position of the engine bleed switches to the "ON" position and the air conditioning packs to the "AUTO" position. Revised checklists were released in June 2005 for the B737-300. This change was incorporated to reinforce correct air conditioning and pressurization system operation.

- Revision of the B737 Flight Crew Training Manual (FCTM) to include a section informing crews on the distinction between the Cabin Altitude Warning and the Takeoff Configuration Warning. This revision had been finalized and approved for publication in May 2005 and was scheduled to be released in the next formal FCTM revision cycle during October 2005. This change was prepared to reinforce correct air conditioning and pressurization system operation as well as to reinforce correct identification of the Cabin Altitude Warning horn.

A post accident review of B737 pressurization incidents from 1998 to 2004 identified 74 incidents in which the cabin altitude warning horn sounded. These 74 incidents included 29 incidents in which the passenger oxygen masks also deployed in the cabin. According to Boeing, the data reviewed showed that the percentage of incidents that involved passenger oxygen system activation declined from 1999-2000 to 2003-2004. The decline suggested that since 2001, crews had more quickly recognized and reacted to pressurization problems and corrected them before the cabin altitude reached 14.000 ft, and this may be attributable, at least in part, to the actions taken by Boeing.

In 2005, the Boeing Company issued a Multi Operator Message (Message Number 1- 116769621-1) in which it informed operators of an October 2005 revision to Boeing 737 Flight Crew Training Manuals to include a new section entitled Air Systems/Cabin Altitude Warning. Among other things, the new section *"... reminds crews how to understand and recognize the differences between cabin altitude and takeoff configuration warnings."*

Chain of Events in Pressurization Incidents and Accidents

A review of pressurization incidents and accidents clearly indicated that aircraft pressurization events (on all B737 but also on other aircraft) had a continuing presence in aviation operations. Multiple issues arose from this review:

- Pre-flight system configuration that could result in inadequate pressurization in flight (e.g. packs or engine bleed selected off, or pressurization mode selector in manual);

- System configuration that persists during subsequent "before start" and after takeoff" checklists. Flight crews' observance of procedures, checklists, and SOPs, particularly during preflight;

- Cabin altitude warning horn not recognized by the flight crew. Flight crews' understanding of the pressurization system, its function, its warning(s), components (switches, placards, annunciations), and consequences of its failures;

- Master Caution and passenger oxygen masks deployment indication not recognized by the flight crew. Pressurization control and indicator design;

- Cabin crew does not advise the flight crew of passenger oxygen masks deployment. Establishing and maintaining open communication between the cabin and the flight deck; and

- Physiological effects of rising cabin altitude not recognized by the flight crew. Flight crews' understanding for and appreciation of gradual cabin depressurization, the insidious effects of hypoxia, and the importance of using supplemental oxygen as a precaution.

It was important to note the recurring nature of many of these issues in incidents and, on occasion, in fatal accidents. It was also relevant to note that the events examined occurred around the world and on a variety of different pressurized aircraft. Many of the events in the incidents, a few of which escalated into fatal accidents, were identical to events that had occurred in the past, despite the issuance of safety recommendations that should have prevented their reoccurrence. The Board was therefore concerned that multiple warnings, and the opportunities they provided for rectification of emerging problems, were not heeded to in a timely fashion and were allowed to lead to further incidents and accidents.

CHAPTER 11

CONCLUSIONS

Flight Crew

1. The flight crew was licensed and qualified for the flight in accordance with applicable regulations.
2. The flight crew held valid medical certificates and was medically fit to operate the flight.
3. Although atherosclerosis was found (minor atherosclerosis for the Captain and extensive atherosclerosis for the First Officer), the Hellenic Air Force Aviation Medical Centre estimated that brain hypoxia was the dominant and determinant cause of incapacitation.
4. The flight crew was adequately rested and their flight and duty times were incompliance with Cyprus DCA and Operator requirements.
5. During the Preflight procedure, the Before Start and the After Takeoff checklists completion, the flight crew did not recognize and correct the incorrect position of the pressurization mode selector (MAN position instead of AUTO).
6. The green light indication that the pressurization mode selector was in MAN (manual) position should have been perceived by the flight crew during preflight, takeoff, and climb.
7. At an aircraft altitude of 12 040 ft and at a cabin pressure that

corresponds to an altitude of 10 000 ft, about 5 minutes after takeoff, the Cabin Altitude Warning horn sounded.

8. The initial actions by the flight crew to disconnect the autopilot, to retard and then again advance the throttles, indicated that it interpreted the warning horn as a Takeoff Configuration Warning.

9. The incorrect interpretation of the reason for the warning horn indicated that the flight crew was not aware of the inadequate pressurization of the aircraft.

10. There were numerous remarks in the last five years by training and check pilots on file for the First Officer referring to checklist discipline and procedural (SOP) difficulties.

11. The flight crew contacted the company Operations Centre Dispatcher and referred to a Takeoff Configuration Warning horn and the Equipment Cooling lights.

12. Communications between the flight crew and the company Operations Centre Dispatcher were not recorded; nor was there a regulatory requirement to record such communications.

13. At an aircraft altitude of 17.000 to 18.000 ft, the Master Caution was activated and was not canceled for 53 seconds. The reason for its activation may have been either the inadequate cooling of the Equipment or the deployment of the oxygen masks in the cabin. The above activation for either of the above two reasons does not permit identification of the other reason. Independently of the Master Caution indication, there are separate indications for both malfunctions on the overhead panel.

14. The flight crew possibly identified the reason for the Master Caution to be only the inadequate cooling of the Equipment that was indicated on the overhead panel, and did not identify the second reason for its activation, i.e., passenger oxygen masks deployment, that was later also indicated on the overhead panel. The crew became preoccupied with the Equipment Cooling fan situation and did not detect the problem with the pressurization system.

15. The workload in the cockpit during the climb was already high and was exacerbated by the loud warning horn that the flight crew did not cancel.
16. The remarks and observations by training pilots and check pilots with respect to the First Officer's performance explained the omissions of the flight crew in it's performance of the Preflight procedures, the Before Start and the After Takeoff checklists, as well as the non-identification of the warnings and reasons for the activations of the warnings on the flight deck during the climb to cruise.
17. Before hypoxia began to affect the flight crew's performance, inadequate CRM contributed to the failure to diagnose the pressurization problem.
18. The flight crew probably lost useful consciousness as a result of hypoxia some time after their last radio communication on the company frequency at 06:20:21 h, approximately 13 minutes after takeoff.
19. Histological examinations revealed the presence of recent myocardial ischemia in both pilots, which according to the Hellenic Air Force Aviation Medical Centre (KAI) was likely due to the extended exposure to hypoxia.
20. The toxicology test measured ethanol (34 mg/dl or 0.034 % weight/volume) in the specimen of the First Officer. The toxicological report stated that in view of the conditions, the finding may have resulted from post-mortem ethanol production.

Cabin Crew

1. The cabin crew members were trained and qualified in accordance with existing regulations.
2. The cabin crew members were adequately rested and their duty times were in accordance with existing regulations.
3. After the deployment of the oxygen masks in the cabin, the cabin crew members would have expected initiation of a descent or at least leveling-off of the aircraft.
4. It could not be determined what actions were taken by the cabin crew members after deployment of the oxygen masks in the

cabin, nor whether any of the cabin crew members attempted to contact the flight crew or enter the flight deck after passenger oxygen masks deployment.

5. Shortly before flameout of the left engine, a member of the cabin crew was observed by an F-16 pilot to enter the flight deck, to sit at the captain's seat, and to attempt to gain control of the aircraft.
6. The above cabin crew member held a Commercial Pilot License.

Aircraft

1. The aircraft held a valid Certificate of Airworthiness.
2. The mass and centre of gravity of the aircraft were within prescribed limits.
3. The aircraft had been supplied with the required amount of fuel. Fuel was not a factor in this accident.
4. No deferred maintenance defects had been recorded.
5. Data retrieved from the non-volatile memory (NVM) of the No. 2 cabin pressurization controller for at least the last 42 flights revealed a pressurization leak or insufficient inflow of air for reasons that could not be determined.
6. There were nine write-ups related to the Equipment Cooling system in the Aircraft Technical Log from 9 June to 13 August 2005.
7. The maintenance actions performed in the early morning hours of the day of the accident comprised:
    - A visual inspection of the rear right door (R2), no defects were found;
    - A pressurization test, no leakage was found.
8. The record of the maintenance actions in the Aircraft Technical Log was incomplete.
9. After the pressurization test, the pressurization mode selector was not selected to AUTO. Although not a formal omission, it would have been prudent to position the pressurization mode selector back to AUTO.
10. The first recorded data of the accident flight on the non-volatile memory (NVM) chip in the cabin pressurization

controller was at 10.000 ft cabin altitude (12.040 ft aircraft altitude). The data showed that the pressurization system was operating in the manual mode.

11. The aircraft departed the holding pattern and started descending from FL340 when the left engine flamed out from fuel depletion. The right engine also flamed out from fuel depletion shortly before impact.

12. The aircraft was structurally intact before impact.

13. The aircraft was destroyed by the impact.

Manufacturer

1. The description in the Boeing AMM for the procedure for the pressurization check (under the heading *"Put the Airplane Back to its Initial Condition"*) was vague. It did not specify an action item that the pressurization mode selector be returned to the AUTO position after the pressurization check.

2. The manufacturer's Preflight procedure and checklists (Before Start and After Takeoff) for checking and verifying the position of controls on the pressurization panel were not consistent with good Human Factors principles and were insufficient to guard against omissions by flight crews.

3. The manufacturer's procedures should have contained enough redundancy to ensure that the pressurization system was properly configured for flight. Because the position of the pressurization mode selector was critical for pressurization, the specific action should have been explicitly listed in the checklists referring to the pressurization system (Before Start and After Takeoff).

4. The use of the same aural warning to signify two different situations (Takeoff Configuration and Cabin Altitude) was not consistent with good Human Factors principles.

5. Over the past several years, numerous incidents had been reported involving confusion between the Takeoff Configuration Warning and Cabin Altitude Warning on the Boeing 737 and NASA's ASRS office had alerted the manufacturer and the aviation industry.

6. Numerous incidents had been reported world-wide involving cabin pressurization problems on the Boeing 737. A number of remedial actions had been taken by the manufacturer since 2000, but the measures taken had been inadequate and ineffective in preventing further similar incidents and accidents.

ATC

1. The air traffic controllers in Nicosia and Athens, who handled flight HCY 522 were properly licensed and properly qualified.
2. The ATC facilities in Nicosia and Athens were appropriately staffed and the communication equipment operated per regulations. There were no communications or navigational aid abnormalities.
3. Nicosia ACC informed by telephone Athinai ACC that flight HCY 522 was not responding to its radio calls while approaching EVENO, but did not use the formal ICAO procedure (Doc 4444) for the two-way Radio Communication Failure (RCF).
4. One minute before the flight entered the Athinai FIR, the Athinai ACC controller "accepted" the flight, but did not seek communication with it when it entered the FIR and failed to contact Athinai ACC as prescribed.
5. The above mentioned actions by Nicosia and Athinai ACCs did not contribute
to the formation of events of the accident.

EASA, JAA and ICAO

1. Despite several EASA, JAA and ICAO audit and follow up audit findings performed on Cyprus DCA, there was no enforcement of implementation of action plans in order to meet its international obligations in the shortest possible time.

Flight HCY522

1. When the flight HCY522 was intercepted by the F-16s, the F-16 lead

pilot reported that there was no visible damage to the Boeing 737 aircraft, that the Captain's seat was vacant, the person in the First Officer's seat was not wearing an oxygen mask and was slumped over the controls, and some seated passengers in the cabin were observed wearing oxygen masks.

2. Shortly before the aircraft started descending, the F-16 pilot reported that a man wearing clothing of a specific color entered the cockpit and sat down in the Captain's vacant seat. He did not appear to be wearing an oxygen mask. He seemed to make efforts to gain control of the aircraft. It was determined that this man was a cabin attendant who held a Commercial Pilot License.

3. When the left engine flamed out due to fuel depletion, the aircraft exited the holding pattern and started a left descending turn, and followed an uneven flight path of fluctuating speeds and altitudes. Shortly before impact, the right engine also flamed out from fuel depletion.

4. The cabin crew member in the cockpit attempted to transmit a MAYDAY message, which was recorded on the CVR. However, the MAYDAY calls were not transmitted over the VHF radio because the microphone key, as shown by the FDR, was not pressed. The performance of the cabin crew member was very likely impaired by the hypoxic and stressful conditions.

5. Three of the four portable oxygen cylinders on board the aircraft had most likely been used.

6. The cabin altitude was calculated to have been about 24.000 ft, while the aircraft was at a cruise level of 34.000 ft (FL340).

7. The duration (30 minutes) of the CVR installed on the aircraft was insufficient to provide key information that would have clarified the chain of events during the climb phase of the flight. The CVR stopped recording when the engines flamed out.

Operator

1. The After Takeoff checklist section referring to the pressurization system in the Operator's QRH had not been updated according to the latest Boeing revision.

2. The manuals, procedures, and training of the Operator, and to a large extent of the international aviation industry, did not address the actions required of cabin crew members when the passenger oxygen masks have deployed in the cabin and, during climb to cruise, the aircraft has not start descending or at least leveled off, and no relevant announcement has been made from the flight deck.

3. The absence of applied hypoxia training at the Operator, and to a large extent at other airlines, for airline transport pilots increased the risk of accidents because of the insidious nature of incapacitation during climb to cruising altitude as a result of pressurization anomalies or gradual loss of pressurization.

4. There were organizational safety deficiencies within the Operator's management structure and safety culture as evidenced by diachronic findings in the audits prior to the accident, including:

   - Inadequate Quality System;
   - Inadequate Operational Management control;
   - Inadequate Quality and Operations Manual;
   - Cases of non-attendance of management personnel at quarterly management quality review meeting, as required;
   - Organization, management, and associated operational supervision not properly matched to the scale and scope of operations;
   - Inadequate monitoring of pilot certificates and training
   - Insufficient involvement of management pilots in managerial duties, due to lack of time;
   - Incompletely updated training and duty records;
   - Lack of updating of some manuals and in part not fully in compliance with regulations;
   - Key management personnel at time performing the work of two positions;
   - Periods of vacant key management positions;
   - Inadequate remedial actions on audit findings, including level one findings, which could cause

- suspension of the AOC.

Cyprus DCA

1. Organizational safety related deficiencies existed within the Cyprus DCA from at least 1999 and continued to the time of the accident, although some corrective actions were exercised since 2003. These deficiencies prevented the DCA from carrying out its safety oversight obligations within Cyprus, as evidenced by findings in previous audits, including:
   - Lack of resources and qualified personnel, and inability to adequately perform the safety oversight activities as required by ICAO;
   - Over-reliance on the UK CAA;
   - Inadequate on-the-job training for Cypriot inspectors to assume the duties for the DCA;
   - Lack of DCA internal expertise to assess the effectiveness or the technical aspects of the UK CAA inspections and the work performed;
   - Ineffectiveness of the DCA in bringing the Cyprus Civil Aviation legislation and regulations into compliance with the international requirements (ICAO Standards and Recommended Practices);
   - Inadequacy of the structure of the DCA to support safety oversight on current and future operations under the present circumstances;
   - No risk management process;
   - Non-exploitation by the DCA of the full scope of contracted services from the UK CAA, related to on-the-job training of Cyprus Flight Inspectors for reasons beyond the control of the UK CAA;
   - Non-assumption of responsibility of the DCA in directing the UK CAA regarding the accomplishment of its contractual duties;
   - Lack of effective implementation of the corrective action plans from previous audits (ICAO - 46.57 %

non-implementation, when an excess of 15% non-implementation generally indicated significant problems in terms of State oversight capability).

Causes

Direct Causes

1. Non-recognition that the cabin pressurization mode selector was in the MAN (manual) position during the performance of the:

   - Preflight procedure;
   - Before Start checklist; and
   - After Takeoff checklist.

2. Non-identification of the warnings and the reasons for the activation of the warnings (cabin altitude warning horn, passenger oxygen masks deployment indication, Master Caution), and continuation of the climb.

3. Incapacitation of the flight crew due to hypoxia, resulting in continuation of the flight via the flight management computer and the autopilot, depletion of the fuel and engine flameout, and impact of the aircraft with the ground.

Latent causes

1. The Operator's deficiencies in organization, quality management and safety culture, documented diachronically as findings in numerous audits.
2. The Regulatory Authority's diachronic inadequate execution of its oversight responsibilities to ensure the safety of operations of the airlines under its supervision and its inadequate responses to findings of deficiencies documented in numerous audits.
3. Inadequate application of Crew Resource Management (CRM) principles by the flight crew.
4. Ineffectiveness and inadequacy of measures taken by the

manufacturer in response to previous pressurization incidents in the particular type of aircraft, both with regard to modifications to aircraft systems as well as to guidance to the crews.

Contributing Factors to the Accident

1. Omission of returning the pressurization mode selector to AUTO after un-scheduled maintenance on the aircraft.
2. Lack of specific procedures (on an international basis) for cabin crew procedures to address the situation of loss of pressurization, passenger oxygen masks deployment, and continuation of the aircraft ascent (climb).
3. Ineffectiveness of international aviation authorities to enforce implementation of corrective action plans after relevant audits.

CHAPTER 12

RECOMMENDATIONS

Safety Actions Taken or in Progress

To NTSB

2005 – 37. On 25 August 2005, the AAIASB recommended to the NTSB that the Boeing Company consider taking action to emphasize flight crew training and awareness in relation to (a) the importance of verifying the bleed and pack system configuration after takeoff and (b) the understanding and recognition of the differences between cabin altitude and takeoff configuration warnings.

Response/Action: On 25 August 2005, the NTSB responded that the Boeing Company was prepared to issue an October 2005 revision to 737- 300/400/500/600/700/800/900/BBJ Flight Crew Training Manuals (FCTM) to include a new section entitled Air Systems/Cabin Altitude Warning reminding flight crews on how to understand and recognize the differences between the two meanings of the warning horn and reminding them of the importance of verifying the bleed and pack system configuration after takeoff.

2005 – 38. On 25 August 2005, the AAIASB recommended

to the NTSB that the Boeing Company clarify the Aircraft Maintenance Manual (AMM) maintenance procedure for Cabin Pressure Leakage Test (05-51-91) to explicitly specify the actions necessary to complete the maintenance test. Currently, under the title of section F *"Put the airplane back to its initial condition"*, there were three action items but none of them referred for the pressure mode selector to be placed in the position AUTO.

Response/Action: On 12 October 2005, the NTSB responded that the Boeing Company had released a Temporary Revision to 737-300/400/500 AMM 05- 5 1-91/20 on 29 September 2005 to include a specific step to put the pressure mode selector in AUTO at the conclusion of the cabin pressure leakage test. This change was planned for official release in the 12 January 2006 revision cycle of that manual. The same change to the corresponding 737- 600/700/800/900/BBJ and 737-100/200 AMM were planned for official releases in the 21 October 2005 and 1 August 2006 revisions, respectively.

2005 – 39. On 13 September 2005, the AAIASB recommended to the NTSB that the Boeing Company consider revising the Aircraft Maintenance Manual (AMM) 05-51-91 by adding an additional step associated with section F *("Put the airplane back to its initial condition")*: to re-install the oxygen mask regulators (if removed) per AMM 35-12-00.

Response/Action: On 1 September, 2006, the FAA responded to the Safety Recommendation. The FAA informed the Board that the Boeing 737 Maintenance Review Board Chairman had advised Boeing of the AMM missing steps in December 2005. Boeing revised the 737-300/400/500 AMM with the 12 January 2006 revision, and added to AMM 05-51-91, paragraph 2.F, a step 4 which stated *"Move the pressurization mode selector on the forward overhead panel to AUTO"*, and a step 5 which stated *"If the crew oxygen mask regulator was removed, then install and test the mask demand regulator (AMM 35-12-86/401)."* As a result, the FAA (Office of

Accident Investigation, Safety Recommendation Review Board) classified the corresponding FAA recommendation as *"Closed – Acceptable Action"*.

2005 – 41. On 23 December 2005, the AAIASB recommended to the NTSB that the Boeing Company consider enhancing the design of the Preflight checklist to better distinguish between items referring to the air conditioning and the pressurization systems of the aircraft and to include an explicit line item instructing flight crews to set the pressurization mode selector to AUTO.

Response/Action: On 10 January 2005, the NTSB responded that the Boeing Company was preparing to issue enhancements to the flight crew procedures associated with the Boeing 737 Cabin Altitude Warning System by issuing a revision to 737-200/300/400/500/600/700/800/900/ BBJ Flight Crew Operations Manuals (FCOM)/Quick Reference Handbooks (QRH). The changes included modification of an existing Normal Checklist (NC), deletion of an existing Non-Normal Checklist (NNC), addition of a new NNC, and change in terminology.

2005 – 42. On 23 December 2005, the AAIASB recommended to the NTSB that the Boeing Company reconsider the design of the Cabin Pressure Control System controls and indicators so as to better attract and retain the flight crew's attention when the pressurization mode selector position is in the MAN (manual) position.

Response/Action: On 30 June 2006, in its comments on the draft Final Report, the Boeing Company responded that a change in the colour of the indicator, as specifically suggested by the AAIASB, could provide a misleading indication to the flight crew that another failure had occurred requiring additional action.

To Cyprus AAIIB

2005 – 40. On 20 October 2005, the AAIASB recommended to the Cyprus Air Accident and Incident Investigation Board that all airlines under the jurisdiction of the Cyprus DCA standardize cabin crew procedures for access to the flight deck and use of the cockpit door, and include relevant training in the Operations Manual.

Response/Action: On 28 November 2005, the Cyprus DCA responded to the AAIASB that such a procedure had been included in Cypriot air operators' manuals already before the Helios Airways accident. After the accident, and still within the framework of national and international regulations, the air operators had made appropriate changes to the procedure for access to the flight deck.

To Hellenic ACC

2005 – 43 On 2 May 2006, the AAIASB recommended to the Hellenic ACC that it consider the need for adding an indication on the label attached to the target of a flight on the radar scope, to draw a controller's attention when radio communication has not been achieved, and that it establish procedures to specify a time limit within which a controller should take the initiative to contact a flight that omitted to report its position when it crossed a compulsory reporting point (FIR boundaries, etc.). ICAO procedures (Doc 4444) stated that action should be taken if a report from an aircraft is not received within a *"reasonable period of time"*, and it is left to regional air navigation agreements to prescribe a specified time interval.

Response/Action: On 28 June 2006, the Hellenic ACC responded that:

- An appropriate procedure has been installed in the software of the radar system, in order to provide a visual indication to the controller if radio communication between the ACC and an aircraft has not been achieved.

- The time limit within which a controller should take the initiative to contact a flight that omitted to report its position has been specified to three minutes and the requirement has been inserted in the ACC Operations Manual.

Actions by the FAA

On 22 June 2006, the FAA issued Airworthiness Directive (AD) 2006-13-13 applicable to all Boeing 737 series. This was an Immediately Adopted Rule, which became effective on 7 July 2006. The AD required revisions to the Airplane Flight Manual (AFM) within 60 days to advise the flight crew of improved procedures for pre-flight setup of the cabin pressurization system, as well as improved procedures for interpreting and responding to the cabin altitude / configuration warning horn.

Specifically, the AD revised the AFM, Normal Procedures, for the Boeing 737 series to include a procedure: *"For normal operations, the pressurization mode selector should be in AUTO prior to takeoff."*

The AD also specified changes to the AFM, Emergency or Non-Normal Procedures sections:

"WARNING HORN–CABIN ALTITUDE OR CONFIGURATION RECALL

Condition: An intermittent or steady warning horn sounds:

- *In flight an intermittent horn indicates the cabin altitude is at or above 10.000 ft;*

- *On the ground an intermittent horn indicates an improper takeoff configuration when advancing thrust levers to takeoff thrust; and*

-
- *In flight a steady horn indicates an improper landing configuration.*

If an intermittent horn sounds in flight:

OXYGEN MASKS AND REGULATORS ON, 100%
CREW COMMUNICATIONS ...ESTABLISH
Do the CABIN ALTITUDE WARNING OR RAPID
DEPRESSURIZATION checklist.

If an intermittent horn sounds on the ground:

   Assure proper airplan takeoffconfiguration.

If a steady horn sounds in flight:

   Assure proper airplane landing configuration.

The FAA advised that once a design change is developed, approved, and available, the FAA may consider additional rulemaking.

Recommended Safety Actions

To EASA/JAA

2006 – 41. EASA/JAA require all airlines to amend cabin crew procedures, so that, when the oxygen masks deploy in the cabin due to loss of cabin pressure or insufficient cabin pressure and if the aircraft does not suspend climb, or level-off or start a descent, the Cabin Chief (or the cabin crew member situated closest to the flight deck) be required to immediately notify the flight crew of the oxygen masks deployment and to confirm that the flight crew have donned their oxygen masks.

2006 – 42. EASA/JAA require aircraft manufacturers to install in newly manufactured aircraft, and on a retrofit basis in older aircraft, in addition to the existing cabin altitude warning horn, a visual and/or an oral alert warning when the cabin altitude exceeds 10.000 ft.

2006 – 44. EASA/JAA require practical hypoxia training as a mandatory part of flight crew and cabin crew training. This training should include the use of recently developed hypoxia training tools that reduce the amount of oxygen a trainee receives while wearing a mask and performing tasks.

To EASA/JAA and ICAO

2006 – 45. EASA/JAA and ICAO require aircraft manufacturers to evaluate the feasibility of installation of a CVR that records the entire flight.

2006 – 46. EASA/JAA and ICAO require all company communications with the aircraft (operations office, technical base/stations, and airport stations) to be recorded.

2006 – 47. EASA/JAA and ICAO require the aircraft manufacturers to also record cabin altitude on the FDR.

2006 – 48 EASA/JAA and ICAO study the feasibility of requiring the installation of crash protected image recorders on the flight deck of commercial aircraft.

2006 – 49. EASA/JAA and ICAO implement a means to record international safety audits of the States' Civil Aviation Authorities, which ensures that the findings can be tracked in depth, action plans are developed and implemented in shortest possible time; and impose the necessary pressure when they become aware that international obligations and standards are not being met by the Authorities.

To The Republic of Cyprus

2006 – 50 The Republic of Cyprus should support by all necessary resources the already under-reorganization Cyprus DCA so that it may be better equipped to carry out the governmental aviation safety oversight functions and to meet its international obligations in the shortest possible time.

CHAPTER 13

IN THE AFTERMATH...

Following the publication of the final report of the crash of Helios Flight 522 a several lawsuits were filed:

1. Families of the dead filed a lawsuit against Boeing on 24 July 2007. Their lawyer, Constantinos Droungas, said *"Boeing put the same alarm in place for two different types of disfunction. One was a minor fault, but the other - the loss of oxygen in the cockpit - is extremely important"*. He also said that similar problems had been encountered before on Boeings in Ireland and Norway. The families are claiming 76 million euros in compensation from Boeing.

2. In early 2008, an Athens prosecutor charged six people with manslaughter. Reports said the suspects were two Britons, one Bulgarian national and three Cypriots.

3. On 23 December 2008, five Helios Airways officials were charged with manslaughter and of causing death by recklessness/negligence. Cyprus' Deputy Attorney General Akis Papasavvas would not identify those charged, but said the suspects had until 26 February (3pm) 2009 to appear before court and answer the charges – effectively to enter a plea.

4. Relatives of the dead filed a class action suit against the Cypriot Goverment – specifically the Department of Civil Aviation for negligence that led to the air disaster. They claim that the DCA was turning a blind eye to airlines' loose enforcement of regulations, and that in general the department cut corners when it came to flight safety.

5. Lieff Cabraser Heimann & Bernstein, LLP, in cooperation with the Cyprus law firm Phoebus, Christos Clerides, N. Pirilides & Associates, of Nicosia and Limassol, announced the other day that families of victims of the crash of Helios Flight 522 filed a lawsuit against The Boeing Company in United States District Court in Chicago, Illinois. One hundred and fifteen passengers and six crew members died on August 14, 2005, when the Boeing 737-200 apparently flew without cabin pressure and then crashed north of Athens, Greece.

GLOSSARY

AAIB	Air Accident Investigation Board (in the United Kingdom)
A/C	Aircraft
AC	Alternate Current
ACC	Area Control Center
AFM	Aeroplane Flight Manual
AFTN	Aeronautical Fixed Telecommunication Network
AFS	Automatic Flight System
A/IR	Class A – Instrument Rating (pilot license)
AMM	Aeroplane Maintenance Manual
AOC	Air Operators Certificate
APP	Approach Control
APU	Auxiliary Power Unit
ATC	Air Traffic Control
ATPL	Airline Transport Pilot License
AWY	Airway
BITE	Built In Test Equipment
CAA	Civil Aviation Authority
CAS	Calibrated Airspeed
CAVOK	Ceiling and visibility OK (No clouds and unlimited visibility)
CG	Center of Gravity
CoA	Certificate of Airworthiness
CoM	Certificate of Maintenance
CoR	Certificate of Registration
CPL	Commercial Pilot License
CPS	Cabin Pressure Controller
CRM	Crew Resource Management
CRS	Certificate of Release to Service
CVR	Cockpit Voice Recorder
CWS	Control Wheel Steering
CPCS	Cabin Pressure Control System
DC	Direct Current
DADC	Digital Air Data Computer
DBA	Deutsche British Airways
DCA	Department of Civil Aviation (in Cyprus)
DCPCS	Digital Cabin Pressure Control System

E	East
E/E	Electrical & Electronic
EASA	European Aviation Safety Agency
ECG	Electrocardiogram
EFIS	Electronic Flight Instrument System
EQA	Equipment Quality Analysis
FAA	Federal Aviation Administration (Civil Aviation Authority in the United States)
FCL	Flight Crew License
FCOM	Flight Crew Operations Manual
FCSU	Flash Card Survivable Store Unit
FCV	Flow Control Valve
FDM	Flight Data Monitoring
FDR	Flight Data Recorder
FIR	Flight Information Region
FL	Flight Level
FMS	Flight Management System
FO	First Officer
ft	Feet
FTL	Flight Time Limitation
FWD	Forward
GPS	Global Positioning System
h	hours
HCAA	Hellenic Civil Aviation Authority
hPa	hectoPascal
ICAO	International Civil Aviation Organization
JAA	Joint Aviation Authorities
JAR	Joint Aviation Requirements
JRCC	Joint Rescue Coordination Center
km	kilometers
kt	knots
LBA	Luftfahrt-Bundesamt (Civil Aviation Authority in Germany)
LED	Light Emission Diode
LoA	Letter of Agreement
LPC	License Proficiency Check
LVO	Low Visibility Operation
MAC	Mean Aerodynamic Cord

METAR	Meteorological Aerodrome Report
MHz	Mega Hertz
MM	Maintenance Manual
N	North
NCO	National Operations Center
NM	nautical miles
NTSB	National Transportation Safety Board (Accident Investigation Authority in the United States)
NVM	Non Volatile Memory
NOSIG	No Significant Change
OFV	Outflow Valve
OLDI	On Line Data Interchange
OM	Operations Manual
OPC	Operator Proficiency Check
PALLAS	Phased Automation of the Hellenic ACC System
PAX	Passengers
PIC	Pilot in Command
PRSOV	Pressure Regulating Shut-Off Valve
psi	Pounds Per Square Inch
psia	Pounds Per Square Inch Absolute
QRH	Quick Reference Handbook
RCF	Radio Communication Failure
ROC	Rate of Climb
RVSM	Reduced Vertical Separation Minima
SFO	Senior First Officer
SLfpm	Sea Level feet per minute
SMC	Stall Management Computer
SOP	Standard Operation Procedures
SSR	Secondary Surveillance Radar
STAR	Standard Instrument Arrival
TRE	Type Rating Examiner
TUC	Time of Useful Consciousness
UTC	Coordinated Universal Time
VHF	Very High Frequency
VMC	Visual Meteorological Conditions
VOR	Very High Frequency Omnidirectional Radio Range
åp	Differential Pressure

Lightning Source UK Ltd.
Milton Keynes UK
04 December 2009